Proceedings in
Information and Communications Technology   6

Yasuhiro Suzuki   Toshiyuki Nakagaki (Eds.)

Natural Computing and Beyond

Winter School Hakodate 2011, Hakodate, Japan,
March 2011 and 6th International Workshop
on Natural Computing, Tokyo, Japan,
March 2012, Proceedings

 Springer

Volume Editors

Yasuhiro Suzuki
Nagoya University
Japan
E-mail: ysuzuki@nagoya-u.jp

Toshiyuki Nakagaki
Future University Hakodate
Japan
E-mail: nakagaki@fun.ac.jp

ISSN 1867-2914
ISBN 978-4-431-54393-0
DOI 10.1007/978-4-431-54394-7
Springer Tokyo Berlin Heidelberg New York

ISSN 1867-2922   (electronic)
ISBN 978-4-431-54394-7   (eBook)

Library of Congress Control Number: 2013933564

CR Subject Classification (1998): J.2, J.3, J.5, I.2, F.4

Typesetting: Camera ready by author and data conversion by Scientific Publishing Services, Chennai, India.

Printed on acid-free paper

Springer is part of Springer Science+Business Media (www.springer.com)

Preface

This book is a joint publication of the Winter School of Hakodate (WSH) 2011 conference at the Future University of Hakodate and the 6th International Workshop on Natural Computing (6th IWNC) at the University of Tokyo. WSH 2011 had been scheduled for March 15–16, 2011, but on March 11, just four days before the event was due to begin, the catastrophic Tohoku earthquake and tsunami struck, causing massive damage to the northeastern coast of Japan.

At the time of the earthquake, one of the co-chairs of WSH 2011, Suzuki Yasuhiro, had been attending a conference in the Tokyo Bay area (at the Nihon Kagaku Miraikan, or National Museum of Emerging Science and Innovation) and was promptly evacuated to a nearby emergency shelter. From the shelter, Prof. Suzuki sent the following email to Prof. T. Nakagaki, the principal organizer of WSH 2011.

From: Suzuki
To: Nakagaki
21:10, March 11, 2011
—
Dear Dr. Nakagaki,
I was in the Tokyo area when the earthquake hit and am now in a nearby shelter.
How is the current situation in Hakodate?
I do not suppose we will have to change the schedule for the winter school.
What do you think?
All the best,
Yasuhiro SUZUKI

As time went by, it became clear that the damage from the tsunami was very serious indeed and that a further disaster, involving the No. 1 power plant at Fukushima, was developing. Almost all airline flights to Japan were cancelled and public transportation in the northeastern and Kanto (Tokyo) regions of Japan were in disarray, making attendance at the WSH impossible for foreign participants. Suzuki and Nakagaki discussed the situation and concluded that it was best to cancel the winter school as it had been planned, but hold a small workshop for whatever participants were able to come to Hakodate:

From: Nakagaki, Suzuki
To: All participants in the Winter School of Hakodate
10:31, March 13, 2011
—

Dear participants in the Winter School of Hakodate,
As all of you know, we have experienced a major catastrophe in Japan and
continue to face unknown circumstances. Here in Hakodate, we have sus-
tained damage near the centre of town, resulting in a suspension of service
for most incoming railways. Fortunately, airways departing and arriving at
the Hakodate airport are nearly on time (with some delays) and municipal
transportation is running as usual.
In light of the larger regional emergency we did discuss outright cancellation
of the Winter School conference, but since the event was scheduled for the
last month of the financial year, full cancellation would likely cause problems
for the management of grants and expenditures. Therefore, we have decided
that it is best to cancel the planned conference but to hold a small workshop
for any participants capable of making the journey to Hakodate as planned.
To participants from abroad we have already announced that the workshop
has been cancelled, as most flights into Japan are suspended and advisories
against travelling to Japan are in effect until the situation stabilizes. If you
cannot attend the smaller workshop due to transportation issues or other
difficulties, please feel free to cancel.
As mentioned, the workshop will be small and will probably take the form
of a casual seminar with a more flexible schedule. We kindly ask that any
participants who have reserved a hotel room near the JR Hakodate station
please confirm their reservations.
We extend our deepest sympathies to those suffering in any way due to the
recent catastrophe and hope that family and friends are safe.
With kindest regards,
Nakagaki and Suzuki

Under these unforeseen circumstances, the workshop proceeded and was at-
tended by a total of seven participants from various parts of Japan. Over the
next few days, we became aware of the massive loss of life due related to the
tsunami, and of the worsening situation at Fukushima. Nevertheless, while in
attendance at the workshop, we tried to focus on the topics at hand: physics,
chemistry, computer science, biology, and aesthetics. We were pleased to find
the discussions intense and energetic, with particular interest focused on Prof.
Akiba's talk on modern arts from the point of view of natural computing. This
talk was based on his book *A New Type of Aesthetics,* which proposed an understanding of aesthetics based on the mechanics of natural algorithms. So well
received was this talk that it inspired the launch of a new research group in
computational aesthetics in the Special Interest Group of Natural Computing
(SIG-NAC), a part of the Japanese Society for Artificial Intelligence (JSAI).

SIG-NAC has been organized by the International Workshop on Natural Computing since 2006.

In the days following the workshop, Tokyo had become something of a "ghost town" due to the lack of electric power. Lighting for commercial uses in train stations and shopping areas was limited, many shops were closed altogether, and a significant number of people were stranded in the city centre.

Almost 12 months passed before we gathered again in Tokyo for the 6th International Workshop on Natural Computing (6th IWNC) at the University of Tokyo, from March 28 to 30, 2012. At this workshop, we were reacquainted with the participants at the WSH in Hakodate, and the computational aesthetics research group convened at a special lunch and symposium.

Because WSH 2011 and 6th IWNC are so closely related, we have decided to edit this special publication, merging papers presented at both the Winter School of Hakodate 2011 and the 6th International Workshop on Natural Computing. The publication includes a wide range of interesting new work.

On the topic of computing with natural media, I. Kunita, S. Sato, T. Saigusa, and T. Nakagaki present "Ethological Response to Periodic Stimulation in *Chara* and *Blepharisma*"; I. Kunita, K. Yoshihara, A. Tero, K. Ito, C. F. Lee, M. D. Fricker, and T. Nakagaki present "Adaptive Path-Finding and Transport Network Formation by the Amoeba-like Organism *Physarum*"; and Y. Fujiwara presents "Aggregate 'Calculation' in Economic Phenomena," illustrating a number of interesting distributions and fluctuations.

In the area of natural computing, M. Hagiya and I. Kawamata present a position paper titled "Towards Co-evolution of Information, Life and Artificial Life"; and Y. Suzuki presents "Harnessing Nature for Computation."

On the topic of computational aesthetics in natural computing, F. Akiba proposes "A Theory of Art Learned from Natural Computing" in which he points out the special significance of natural computing when considering computational aesthetics; M. Goan, K. Tsujita, T. Ishikawa, S. Takashima, S. Kihara, and K. Okazaki present the "Asynchronous Coordination of Plural Algorithms and Disconnected Logical Types in Ambient Space"; and J. Watanabe offers "Aesthetic Aspects of Technology-Mediated Self-Awareness Experience" along with several pieces of related artwork.

On the topic of synthetic biology in natural computing, N. Noman, L. Palafox, and Hitoshi Iba propose a "Method for the Reconstruction of Gene Regulatory Networks from Gene Expression Data Using a Decoupled Recurrent Neural Network Model"; L. Palafox, N. Noman, and H. Iba investigate the use of "Evolutionary Techniques for Inference in Gene Regulatory Networks"; and R. Sekine and M. Yamamura review the "Design and Control of Synthetic Biological Systems."

We sincerely thank all contributors for their interesting work and their prompt support in editing this joint volume. We express special thanks to Prof. Masami Hagiya from the University of Tokyo on organizing 6th IWNC and A. Hofmann from Springer, Heidelberg, and to the staff at Springer Japan for this special publication. WSH 2011, 6th IWNC, and this publication were supported by

Grant-in-Aid for Scientific Research on Innovative Areas No. 23119008 "Synthetic Biology for the Comprehension of Biomolecular Networks" and No. 24104002 "Molecular Robotics" and Grant-in-Aid for Scientific Research (B) No. 23300317 and (C) No. 24530106.

December 2012

Yasuhiro Suzuki
Toshiyuki Nakagaki
Co-Chairs WSH2011 and 6th IWNC

Organization

WSH 2011 and 6th IWNC were organized by the Special Interest Group of Natural Computing (SIGNAC) in the Japanese Society for Artificial Intelligence. The 6th IWNC was supported by Scientific Research on Innovative Areas "Synthetic Biology for the Comprehension of Biomolecular Networks", Grant-in-Aid for Scientific Research on Innovative Areas.

Program Committee

Conference Chair of WSH2011:

Toshiyuki Nakagaki   Future University Hakodate, Japan
Yasuhiro Suzuki   Nagoya University, Japan

Conference Chair of 6th IWNC:

Yasuhiro Suzuki   Nagoya University, Japan

Program Committee Members

Fuminori Akiba   Nagoya University (Japan)
Daniela Besozzi   University of Milan, Bicocca (Italy)
Alberto Castellini   University of Verona (Italy)
Taichi Haruna   Kobe University (Japan)
Hiroyuki Kitahata   Chiba University (Japan)
Satoshi Kobayashi   University of Electoro-Communication (Japan)
Vincenzo Manca   University of Verona (Italy)
Masami Hagiya   University of Tokyo (Japan)
Giancarlo Mauri   University of Milan, Bicocca (Italy)
Marion Oswald   Vienna University of Technology (Austria)
Sigeru Sakurazawa   Future University, Hakodate (Japan)
Junji Watanabe   NTT Communication Science Laboratories
                              (Japan)
Nozomu Yachie   University of Toronto (Canada)
Takashi Yokomori   Waseda University (Japan)

Table of Contents

Natural Computing

Ethological Response to Periodic Stimulation in *Chara* and
Blepharisma.. 3
    Itsuki Kunita, Sho Sato, Tetsu Saigusa, and Toshiyuki Nakagaki

Adaptive Path-Finding and Transport Network Formation by the
Amoeba-Like Organism *Physarum* 14
    Itsuki Kunita, Kazunori Yoshihara, Atsushi Tero, Kentaro Ito,
    Chiu Fan Lee, Mark D. Fricker, and Toshiyuki Nakagaki

Aggregate "Calculation" in Economic Phenomena: Distributions and
Fluctuations.. 30
    Yoshi Fujiwara

Towards Co-evolution of Information, Life and Artificial Life 39
    Masami Hagiya and Ibuki Kawamata

Harness the Nature for Computation 49
    Yasuhiro Suzuki

Things Theory of Art Should Learn from Natural Computing.......... 71
    Fuminori Akiba

Study on the Use of Evolutionary Techniques for Inference in Gene
Regulatory Networks 82
    Leon Palafox, Nasimul Noman, and Hitoshi Iba

Reconstruction of Gene Regulatory Networks from Gene Expression
Data Using Decoupled Recurrent Neural Network Model 93
    Nasimul Noman, Leon Palafox, and Hitoshi Iba

Design and Control of Synthetic Biological Systems 104
    Ryoji Sekine and Masayuki Yamamura

Satellite Symposium on Computational Aesthetics

Preface: Natural Computing and Computational Aesthetics 117
    Fuminori Akiba

The Significance of Natural Computing for Considering Computational
Aesthetics of Nature.. 119
    Fuminori Akiba

Perceiving the Gap: Asynchronous Coordination of Plural
Algorithms and Disconnected Logical Types in Ambient Space......... 130
    Miki Goan, Katsuyoshi Tsujita, Takuma Ishikawa,
    Shinichi Takashima, Susumu Kihara, and Kenjiro Okazaki

Aesthetic Aspects of Technology-Mediated Self-awareness
Experiences 148
    Junji Watanabe

Author Index... 155

Part I
Natural Computing

Ethological Response to Periodic Stimulation
in *Chara* and *Blepharisma*

Itsuki Kunita[1], Sho Sato[2], Tetsu Saigusa[3,4], and Toshiyuki Nakagaki[1,5]

[1] Future University Hakodate, 116-2 Kamedanakano-cho,
Hakodate, Hokkaido, Japan 041-8655
[2] School of Science, Department of Biological Sciences,
Hokkaido University, Sapporo, Hokkaido, Japan 060-0810
[3] Department of Clinical Chemistry and Laboratory Medicine,
Graduate School of Medical Sciences, Kyushu University, Fukuoka, Japan 812-8582
[4] Research on Dementia, Japan Foundation for Aging and Health,
Chita, Aichi, Japan 470-2101
[5] JST, CREST, 5, Sanbancho, Chiyoda-ku, Tokyo, Japan 102-0075

Abstract. To study how organism responds to periodic stimulation is meaningful since it may be an approach to an elementary capacity of time memory and learning in chronological events. We reported here that the ability of time memory found in true slime mold *Physarum* was also found in a protozoan ciliate, *Blepharisma japonicum* and a green plant *Chara*. Stimulation of temperature or light was repeated several times in a regular period, and the creature anticipated the next timing of stimulation. After the anticipatory behavior disappeared some time later, another single stimulation triggered recalling of periodicity of the previous stimulation. We discuss that the observed capacity is expected to be common in a range of species as the similar capacity has been reported in true slime mold *Physarum*. The observed responses were, however, dependent of individual of organism and a wide range of different responses was observed. We need an extensive study of both experimental characterization and mathematical modeling of ethological dynamics. *abstract environment.*

Keywords: *Physarum*, cell memory, subcellular computing, primitive intelligence.

1 Introduction

Organisms in the wild life are exposed to various kinds of stimulation from their environments. Stimulations are not merely single but repeated many times in time sequence[1–4]. The time sequence can be in general described by a function of time. Such a function is properly approximated by a finite Fourier series of trigonometric functions. To study response to stimulation of regular single frequency is elementary since response to each single frequency contributes to organization of complex behavior induced by realistic time sequence of stimulation in the wild. Although the original time sequence is much more complicated,

Y. Suzuki and T. Nakagaki (Eds.): WSH 2011 and IWNC 2012, PICT 6, pp. 3–13, 2013.

here we study the elementary character of organism behavior in response to regularly periodic stimulation.

In 2008, an interesting response to a periodic stimulation was reported in a true slime mold, plasmodium of *Physarum polycephalum*. Plasmodium showed a kind of time memory[5]. Stimulation of low temperature and low humidity were applied three times in a regular period, and the creature anticipated the next timing of stimulation. After the anticipatory behavior disappeared some time later, another single stimulation triggered recalling of periodicity of the previous stimulation. The authors proposed a possible mechanism of observed behavior. An idea to be emphasized is that these behaviors can be realized in dynamics for collective motion of intracellular chemical oscillators. The model equation is simple and generic. This finding may give a hint at evolutionary origin of capacity of time memory[6].

However, it is unclear whether the ability of time memory is specific to *Physarum* only or common to a range of different species. If it is common, the capacity is expected to be general more or less. So seeking for the similar ability in a different organism from *Physarum* is meaningful. We reported here the similar ability of time memory in a protozoan ciliate, *Blepharisma japonicum* and a green algae *Chara*.

In this study, in addition to involvement of those different species, three kinds of physical nature of stimulation, temperature, light and electric current, were tested. If the similar capacity is observed independently from the difference of physical nature of stimulation, the capacity can be organized somewhat in a central unit of information processing, rather than peripheral sensory activity.

Discussion was made on general features that was observed among species and among three kinds of stimulation. Lastly we examined if the model equation proposed in the previous paper[5] was still applicable to *Blepharisma* and *Chara*.

2   Method

2.1   Organism and Culture

A protozoan ciliate, *Blepharisma Joponicum* (wild type), was purchased from Kyoto Kagaku Co. (Kyoto, Japan). The ciliate was cultured at 24 ^{o}C in dark, in a Petri-dish (12 cm in the diameter) with the culture medium of 100 times diluted Chokley solution (0.1 g/l NaCl, 0.004 g/l KCl, 0.006 g/l $CaCl_2$, in the final concentration) and several grains of rice. As the medium was not sterilized, some bacteria and protozoa co-existed. About twenty to thirty organisms were collected from the culture dish, and put them in the smaller arena with disk-shape (10 mm in the diameter and 1 mm in the thickness), which was served for experimental observation. The arena was immersed in the water bath (24 ^{o}C in dark) and fixed in order to keep the temperature constant and stable (see Fig. 1a).

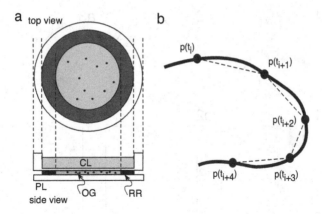

Fig. 1. Experimental methods for *Blepharisma*. (a) Schematic illustration for top and side views of experimental arena in order to observe swimming behavior in *Blepharisma*. The organisms were allowed to swim in the arena with a disc shape (diameter: 10 mm, height: 1 mm). In order to change temperature of water in the arena, coolant was poured on the arena but the coolant was isolated from water of arena with a thin plastic. PL: plastic plate, CL: coolant, OG: organism in the arena, RR: rubber ring. (b) Characterization of swimming behavior. From a real trajectory of swimming (thick sold line), the position of organism, $p(t_i)$, was measured every one second. Swimming distance was approximated by the straight distance every one second, $|p(t_i) - p(t_{i+1})|$ as indicated by the dashed line.

2.2 Observation of Swimming Behavior in Response to Periodic Stimulation in *Blepharisma*

The arena was put on the stage for observing behaviors of free swimming by the CCD video camera, under the infra-red light only. The video camera setup was placed in the incubator, in which temperature was controlled and fixed at 24 $^\circ C$. The video image was recorded and served for image analysis.

After the ciliates in the arena were left gentle on the observation stage without any agitation for a few hours, swimming behavior was recorded. Two kinds of stimulation were applied. One was the stimulation of low temperature. The procedures were the following. The 24 $^\circ C$-water in the water bath was replaced by the $-18^\circ C$-coolant (saturated NaCl aqueous solution), and it was kept for five minutes and 24 $^\circ C$-water was set back again. This was the low temperature stimulation. The other was stimulation of white-light illumination with the light source of fluorescent tube (20 lux).

One of the two types of stimulation was repeated multiple times in a regular interval of time, and a single stimulation was applied again after a long time was passed enough for behavioral response to the previous periodic stimulation to be negligible.

The trajectory of free swimming was traced at the video rate by the computer software, previously developed[7]. Along the trajectory, we picked up the position of creature every one second, and the position was expressed by $p(t_i)$ at time

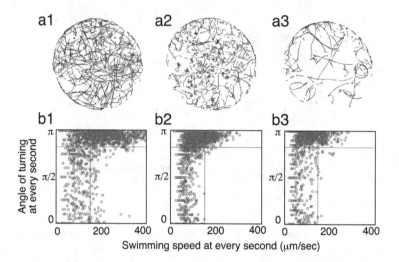

Fig. 2. Typical trajectories of swimming in *Blepharisma*. Upper panels are superimposed four-second-long trajectories in the observation arena. Lower panels are characterization of swimming trajectory, measured by swimming speed and angle of turning at every second. (a) in normal conditions before the cold stimulation, (b) in cold stimulation of PS_n, (c) in the anticipatory occasion of A_n. Swimming was slower and the direction often turned in the cold stimulation. Similar tendency was observed in the anticipatory occasion although the response was much weaker. This tendency was quantitated by the quantity DS/RG (see the method in detail).

t_i discretized in second. The trajectory was approximated by a straight line segment that connected two positions of $p(t_i)$ and $p(t_{i+1})$. Thus the original trajectory was replaced by combination of line segments (see Figs. 1b,2).

Every four second, the distance of swimming (DS) was calculated as

$$DS(t_i) = \sum_{j=0}^{3} |p(t_{i+j}) - p(t_{i+j+1})|.$$

We defined the range (RG) in order to characterize spatial extent of swimming trajectory as

$$RG(t_i) = \max |p(t_{i+j}) - p(t_{i+k})|,$$

where $j, k = 0, 1, 2, 3, 4$. RG takes a large and a low values when the swimming trajectory is straight and localized, respectively. In other words, RG is the largest distance between any two positions p in the period of four seconds. The ratio of two measures, DS/RG, is a non-dimensional quantity, which indicates features of swimming behavior. This index DS/RG is larger as the ciliates turns more frequently. The minimum value is one when it swims straight.

2.3 Measurement for Slowdown in Protoplasmic Flow, Induced by Periodic Stimulation in *Chara*

Chara, a green plant in fresh water like pond and river, was cultured in natural pond water with natural pond soil in 20 to 23 °C. A main stem of the plant with about five nodes was cut and served for experiment. A whole piece of the cut stem was exposed to an external stimulation (see Fig. 3).

Two kinds of stimulation were applied. One was the stimulation of low temperature. The procedures were the following. The specimen was soaked into water (23 °C) in a petri dish. The 23 °C-water in the dish was replaced by the 0°C-water, and it was kept for one minute and 23 °C-water was set back again. This was a single stimulation of cold water. This stimulation was repeated periodically.

The other was stimulation of electric current through the stem of specimen. Voltage of 1.5 volt was loaded with a resistor 60 Ω that was connected parallel to the specimen (see Fig.1c). Period of load lasted for one minute. This was the single stimulation and it was repeated periodically.

Protoplasmic flow was observed in a middle inter-nodal cell in the specimen under the microscope. Image of flow was recorded by the video-camera. Intracellular particles that flew in the cell were tracked by the personal computer and the

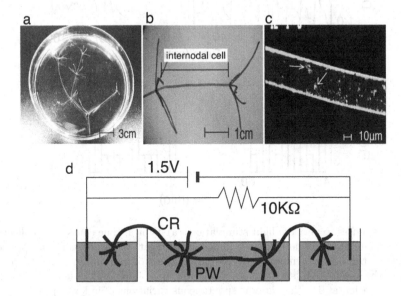

Fig. 3. Pictures of *Chara*. (a) Appearance of green stem of plant in the petri dish. (b) Inter-nodal cell in the stem. (c) Some flowing particles (indicated by arrows) in the inter-nodal cell. (d) Schematic illustration for periodic stimulation of electric current. CR: specimen, PW: pond water in dish. The specimen was soaked into three containers of left, center and right as connecting two containers of left and center, and center and right through the stem of plant. An inter-nodal cell in the center container was observed by a microscope.

originally developing software (particle tracking velocimetry). Flow velocity was calculated and averaged every 10 seconds. The original time variations of flow velocity was smoothed by the moving average (time window was two minutes).

3   Results

3.1   Spontaneous Increases of Swimming Measure at an Anticipatory and a Recalling Occasions in *Blepharisma*

Figure 4 shows a typical response to the repetitive stimulation of light (one minutes) once every five minutes. The occasions of stimulation were indicated by $PS1, PS2, \cdots, PS10, TR$ and the behavior measure DS/RG increased by the stimulation. This means that the organism tends to frequently turn the swimming direction. After the stimulation, some spontaneous increases were periodically observed

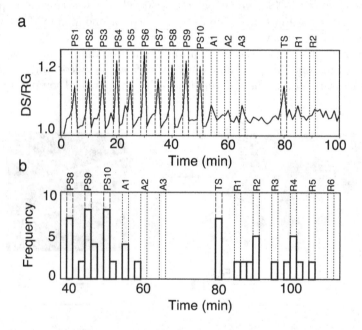

Fig. 4. Responses to periodic light stimulation. (a) Typical time course in a collection of approximate thirty organisms in a single experiment. In response to the periodic stimulation, the behavior index (DS/RG, straightness of swimming) went down and up. After the stimulation, spontaneous down and up of index was observed. After the single stimulation at TS, the periodic spontaneous variations of index appeared again. PSn: the occasion of nth stimulation, An: the nth occasion of anticipatory response, TS: the occasion of single stimulation several hours after the periodic stimulation, Rn: the nth occasion of recalling response. (b) Statistical occurrence of spontaneous decrease in the swimming index with respect to time. The frequency was clearly higher at A1 than at the neighbors of A1. After the stimulation at TS, the frequency was higher at R2, R3 and R4 than at their neighbors. Anticipatory and recalling behaviors were observed although the responses were weak. N=8.

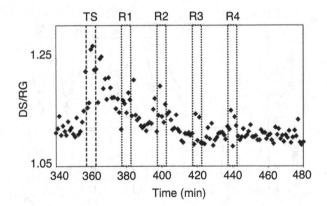

Fig. 5. Recalling responses to repetitive cold-stimulation in six times with the period of 20 minutes. Time courses were superimposed from four experiments. In each experiment, the time course was obtained from a collection of about thirty organisms. Almost 120 organisms were observed for this figure. Six hours after six times of stimulation, the single stimulation was given at TS. Significant increase of measure was observed every occasion of R1 to R4.

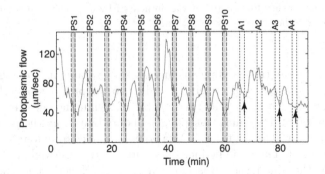

Fig. 6. Typical time course of protoplasmic streaming in response to the periodic electric stimulation. The speed of streaming slowed down at every occasion of stimulation (PS1 to PS10). Spontaneous slowdown was often observed just at the occasions of An (indicated by the arrows).

and the periodicity seemed similar to that of experienced stimulation (see the time section from A1 to A3), loosely speaking. The spontaneous increases happened just at A1, A2 and A3. Once this response of spontaneous increase disappeared around 70 min, the single stimulation was applied at TS.

Figure 4b shows the statistical occurrence of increase (N=8). The spontaneous increases were significant at A1 although environmental conditions were kept constant. After the single stimulation at TS, the periodic spontaneous increases clearly appeared again at R2, R3, R4 and R5 in the similar period to the previously experienced one although the single stimulation had no information

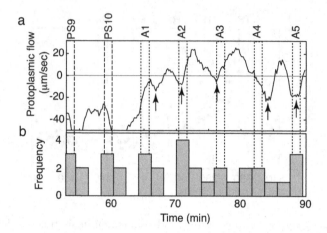

Fig. 7. Statistical examination of responses to the electric stimulation. (a) the flow speed averaged over five repeats. The speed shown in Figure was the difference from the time average of streaming over an entire experiment, so that deviations of flow speed in five organisms were canceled out. The spontaneous slowdown appeared around the occasion of An (indicated by the arrows). (b) Statistical frequency of spontaneous slowdown. A typical slowdown counted was indicated by the arrows in Figure a. High frequency was periodically observed at A1 and A2. N=5.

of periodicity. We say the spontaneous increases at An and at Rn 'anticipatory' and 'recalling' behavior, respectively, according to the previous paper[5].

Figure 5 shows the statistical time course of swimming measure after the single stimulation at TS, which was given six hours later from the periodic cold stimulation (period 20 min, 6 times). Data points obtained from four experiments were superimposed. The periodic increases were clearly observed at R1, R2, R3 and R4 while the stimulation was not given at all. Nonetheless, anticipatory response was not observed in the case.

3.2 Spontaneous Slowdown of Protoplasmic Flow at an Anticipatory Occasions in *Chara*

Figure 6 shows a typical time course of protoplasmic flow in response to the periodic stimulation of electric current (period 6 min, 10 times). The flow speed decreased with the stimulation and the spontaneous slowdown took place periodically at An (indicated by arrows) although the slowdown was sometimes not so clear at A2. Figure 7 shows statistical confirmation of anticipatory responses by measuring average speed of protoplasmic flow (a) and statistical occurrence of slowdown (b). The spontaneous slowdown was clearly periodic and corresponded to the occasion of An (indicated by the arrows) although there sometimes was a slight shift of time like A1 and A4. In the frequency, anticipatory response was significant at A1 and A2.

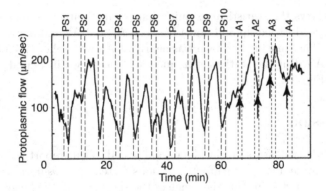

Fig. 8. Typical time course of protoplasmic streaming in response to the periodic cold stimulation with a period of 6 minutes. The speed of streaming slowed down at every occasion of stimulation (PS1 to PS10). Spontaneous slowdown was often observed just at the occasions of An (indicated by the arrows).

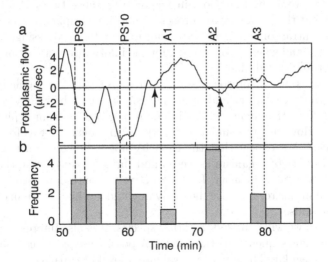

Fig. 9. Statistical examination of responses to the cold stimulation. (a) the flow speed averaged over five repeats. The speed shown in Figure was the difference from the time average of streaming over an entire experiment. The spontaneous slowdown appeared around the occasion of An (indicated by the arrows). (b) Statistical frequency of spontaneous slowdown. A typical slowdown counted was indicated by the arrows in Figure a. High frequency was periodically observed at A1 (weak), A2 (strong) and A3 (intermediate). N=5.

In the periodic cold stimulation (period 6min, 10 times), the response was similar to that in the electric stimulation as the typical time course was shown in Fig. 8. The protoplasmic flow spontaneously slowed down around the time of An, similarly to the response induced by the cold stimulation. The slowdown was, however, sometimes not clear at A1, for instance.

Figure 9 shows the statistical examination of anticipatory response. The averaged flow speed varied up and down between A1 and A2 and between A2 and A3 (Fig. 9a). The time of slowdown was nearly at An while there might be slight shift to just An. But the statistical frequency of slowdown increased just at An as shown in Fig. 9b.

4   Discussion

On the periodicity of stimulation, the period tested in this report was limited at 5, 6 and 20 min and shorter while the periods tested in *Physarum* were 30, 40, 50, 60, 70, 80 and 90 min. But the range of the shortest and the longest periods, $20/5(=4)$, was rather wider than that of $90/30(=3)$. In such new periods tested in the report, anticipatory and/or recalling behaviors are observed.

In this report, not only temperature stimulation but also light and electric stimulation were newly tested. The organisms tested were capable of the anticipatory and/ or recalling behaviors in response to those types of stimulation. This implies that the behaviors are processed not in a peripheral sensory system but in rather central (or common) system for information processing. A question arise then: what and where is the system for the information processing and how it works in Protozoa and plant?

Blepharisma is one of protozoa similar to *Physarum*. The capacity of time memory like the anticipatory and recalling behaviors may be widely common in protozoa. On the other hand, *Chara* is far from protozoa from a phylogenetic point of view. However, the similar capacity of time memory was found. What does this mean? We expect that the capacity might be common much widely in phylogenetic tree of organism as we remind the other suggestion that human can also have a similar capacity. It is attractive to continue to study the time memory found in this report since the study might give a hint at origin of memory capacity in organism.

In the future, we will examine the other species. It is also interesting to test a more complex time sequence of stimulation. A possible sequence can be multiply periodic and irregular. Then we may evaluate a level of capacity.

Because the responses seems not so simple as observed in this report, we need a powerful method to analyze oscillatory behaviors in general. Frequency analysis and mode analysis are of strong interest. Oscillatory behaviors in *Physarum* have been widely studied experimentally and theoretically from a dynamical system point of view[5, 8–14]. The dynamical system point of view is to be helpful as well. Namely it is promising to compare to dynamic behaviors of visco-elastic mechanics and biochemical reaction kinetics in response to external periodic forcing.

Acknowledgements. This research was supported by JSPS KAKENHI 20300105 and by Strategic Japanese-Swedish Research Cooperative Program, Japan Science and Technology Agency (JST).

References

1. Church, R.M.: The internal clock. In: Hulse, S.H., Fowler, H., Honing, W.K. (eds.) Cognitive Processes in Animal Behavior. Erlbaum, Hillsdale (1978)
2. Roberts, S., Church, R.M.: Control of an internal clock. Journal of Experimental Psychology. Animal Behavior Process 4, 318–337 (1978)
3. Meck, W.H., Church, R.M., Olton, D.S.: Hippocampus, time, and memory. Behavioral Neuroscience 98, 3–22 (1984)
4. Gould, J.L., Gold, C.G.: The Honey Bee (Scientific American Library: 186). W.H. Freeman and Company, San Francisco (1988)
5. Saigusa, T., Tero, A., Nakagaki, T., Kuramoto, Y.: Amoebae anticipate periodic events. Physical Review Letters 100(1), 08101 (2008)
6. Ball, P.: Cellular memory hints at the origins of intelligence. Nature, News 451, 358 (2008)
7. Matsumoto, K., Takagi, S., Nakagaki, T.: Locomotive Mechanism of Physarum Plasmodia based on Spatiotemporal Analysis of Protoplasmic Streaming. Biophysical Journal 94, 2492–2504 (2008)
8. A mathematical model for period-memorizing behavior in Physarum plasmodium. J. Theor. Biol. 263, 449–454 (2010)
9. Frequency Coupling Model for Dynamics of Responses to Stimuli in Plasmodium of Physarum polycephalum. J. Phys. Soc. Japan. 66, 1638–1646 (1997)
10. Coggin, S.J., Pazun, J.L.: Dynamic complexity in *Physarum polycephalum* shuttle streaming. Protoplasma 194, 243–249 (1996)
11. Kakiuchi, Y., Ueda, T.: Multiple oscillations in changing cell shape by the plasmodium of *Physarum polycephalum*: general formula governing oscillatory phenomena by the *Physarum* plasmodium. Biol. Rhythms Res. 37, 137–146 (2005)
12. Winfree, A.: The Geometry of Biological Time, 2nd edn. Springer, New York (2001)
13. Kuramoto, Y.: Chemical Oscillations, Waves, and Turbulence. Springer, Heidelberg (1984)
14. Ueda, K., Takagi, S., Nishiura, Y., Nakagaki, T.: Mathematical model for contemplative amoeboid locomotion. Physical Review E 83, 021916 (2011)

Adaptive Path-Finding
and Transport Network Formation
by the Amoeba-Like Organism *Physarum*

Itsuki Kunita[1], Kazunori Yoshihara[1], Atsushi Tero[2], Kentaro Ito[3],
Chiu Fan Lee[4], Mark D. Fricker[5], and Toshiyuki Nakagaki[1,6]

[1] Department of Complex and Intelligent Systems,
Faculty of Systems Information Science, Future University of Hakodate,
116-2 Kamedanakano-cho, Hakodate, Hokkaido, Japan 041-8655
[2] Institute of Mathematics for Industry, Kyushu University,
744 Motooka, Nishi-ku, Fukuoka 819-0395, Japan
[3] Department of Mathematical and Life Sciences,
Faculty of Science, Hiroshima University
[4] Department of Bioengineering,
Imperial College London, London, SW7 2AZ, UK
[5] Department of Plant Science, University of Oxford,
Parks Road, Oxford, OX1 3RB, UK
[6] JST, CREST, 5, Sanbancho, Chiyoda-ku, Tokyo, Japan 102-0075
nakagaki@fun.ac.jp

Abstract. The giant amoeba-like plasmodia of *Physarum* is able to
solve the shortest path through a maze and construct near optimal
functional networks between multiple, spatially distributed food-sources.
These phenomena are interesting as they provide clues to potential biological computational algorithms that operate in a de-centralized, single-
celled system. We report here some factors that can affect path-finding
through networks. These findings help us to understand more generally
how the organism tries to establish an optimal set of paths in more complex environments and how this behaviour can be captured in relatively
simple algorithms.

Keywords: *Physarum*, combinatorial optimization, subcellular computing, primitive intelligence.

1   Introduction

The plasmodium of the true slime mold *Physarum polycephalum* grows as a
giant, acellular amoeba, which transports and circulates nutrients and signals
through network structure of pulsating tubular veins. The network can reorga-
nize within hours in response to external stimulation and environmental changes.
The first evidence that *Physarum* could connect multiple separate food-sources
(FSs) through a tubular network was reported in 1996[1]. The author described
that the shape of network seemed to be optimal or quasi-optimal in some physiological sense, and inferred that this might reflect the operation of a biological

Y. Suzuki and T. Nakagaki (Eds.): WSH 2011 and IWNC 2012, PICT 6, pp. 14–29, 2013.

algorithm to compute the optimal network solution. Thus studying network development in organisms such as *Physarum* might provide the key to understanding how information processing can be performed in cellular system, heralding the advent of bio-inspired computation.

In 2000, *Physarum* was shown to be able to solve the shortest path through a maze [2–4]. When the organism was allowed to completely fill a maze with two FSs placed at the exits, it rapidly migrated towards the FSs and colonized them, leaving a single thick tube connecting the FSs by the shortest path. This adaptive re-distribution of it's cell body enabled the organism to maximize contact with the FSs, to absorb nutrients as fast as possible, whilst retaining connectivity and communication throughout it's length. A simple mathematical model was proposed to capture this behaviour, based on physiological mechanism linking adaptive development of each tubular element to the mass flow passing through the tube itself[5, 6]. At this stage, the experiments showed that the plasmodium was able to distinguish between paths of relatively similar length, but the level of discrimination was not rigorously determined.

In 2004, the computational ability to find a wide range of near-optimal solutions was reported for networks connecting many FSs distributed in a plane space [7, 8]. The experimental systems represent multi-objective optimization problems, and the solutions found by *Physarum* approximated a Pareto frontier in the space spanned by multiple evaluation-functions.

Given that the computational ability of *Physarum* to solve such geometric problems could be expressed in relatively simple mathematical model, a new method for designing a social transport network inspired by *Physarum*-type algorithms was proposed at the international conference on Dynamical Systems organized by Society of Industrial and Applied Mathematics (SIAM) in May 2005. As an concrete example, the ability of *Physarum* to design a railway network around Tokyo was examined, using the distribution of FSs to reflect the geographical pattern of major cities in Tokyo region. *Physarum* was able to make a multi-functional network that connected all the locations with a balance between the total length of the network, the shortest distance between "cities" and the level of fault tolerance to disconnection. Such *Physarum*-type algorithms were also applied to other graph problems, including navigation through a road networks in response to spatio-temporal variations of congestion, and the problem of finding the Steiner Minimum Tree across a network [9–13]. Physarum computing thus represents a new method of bio-inspired problem-solving and has attracted considerable attention from engineers, scientists and the wider community.

Despite this promising start, only a limited range of experimental maze and network configurations have been tested so far, and our current knowledge of the complex behaviour of this fascinating organisms is still limited. This report examines adaptive responses to more complex, but well specified geometrical mazes and network challenges. In particular, we start to explore configurations where the plasmodium was challenged with multiple parallel routes that differed in length and tortuosity (Fig. 1).

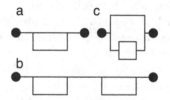

Fig. 1. Schematic diagram showing the topological shapes of mazes used. (a, b) represent the layout used previously and focus on binary decisions based solely on the shortest path length. This paper also explores a new topological configuration with multiple parallel paths that vary in length and tortuosity (c).

2   Material and Methods

2.1   Preparation of Organism and Maze

The organism we used for experiments was the plasmodium of the true slime mold *Physarum polycephalum*. Mazes were made as described previously [3]. Previous mazes followed the layout shown in Fig 1a and 1b. Here we tested different topological layouts with multiple parallel paths (see Fig 2a,b). This gave control of the difference in overall length and tortuosity between the different paths. Two configurations were tested: in both shapes, two food-sources were connected through three paths indicated by numbers a_1, a_2 and a_3. While the path a_3 was single all the way, paths a_1 and a_2 ran parallel to each other over part or all of the route. By varying the location of the FS, indicated by black dots, the relative length of path a_3 was varied with respect to a_2 and a_3. The length difference, termed α, was given by (length of a_1)/(length of a_3), where the lengths of a_1 and a_2 were always the same. It should also be noted that the actual conductivity through tubes of different lengths was also strongly dependent on their thickness, as conductivity follows the Hagen-Poisueille flow approximation. In Fig. 2a, the number of turning points varied between the different paths, whilst in fig. 2b, the number of turning points was the same. In addition, two dead-end paths were included to ensure that the same initial volume of plasmodium was present in each arm of the maze at the start.

2.2   Comparison of Network Structure between Real Transport Networks in the Hokkaido Region, Japan, and *Physarum* Networks Connecting FSs

Twenty-four FSs were set out to represent the geographical locations of the major towns in the Hokkaido region of northern part of Japan (see Fig. 5), following the method described previously [11].

Three recipes were used to prepare the FSs: (1) each FS was formed from a small portion of kneaded dough that included ground oat flake powder and water only, with all FSs being equal in size (see Fig. 5a); (2) FSs were made using the

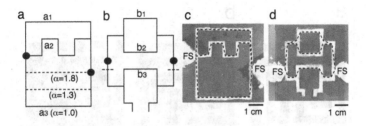

Fig. 2. Two different realizations of the topological layout shown in Fig 1c are shown in (a) and (b) with representative images of the experimental systems showing the typical movement of the *Physarum* plasmodium shown in (c) and (d), respectively. In (a) the maze has two parallel paths (a_1, a_2), that are always the same length, and a third, single path (a_3). When two FSs were placed at the locations indicated by black dots in the figure, the length of these three paths is the same (provided the width of the paths is ignored). By varying the position of the FSs, indicated by the dashed lines, the relative length of path a_3 with respect to a_1 and a_2 could be varied. The length difference was characterized as $\alpha = $ (length of a_1)/(length of a_3). The path a_2 has 7 turning points, whilst the other two paths have only two. In (b) the maze has the same topology as (a), but the number of turning points is the same in all paths. In addition, to make the initial distribution of the organism the same in all paths, two dead-end paths were included in the lower arm of the maze. (c) shows the distribution of the organism completely filling the maze just after presentation of food-sources (FS) in configuration (a). (d) shows the result of an experiment with configuration (b) after several hours when the organism has accumulated at the FSs and some of the connection paths disappeared. The shortest path remained through the lower arm of the maze, but sometimes additional routes still remained. Scale bar in (c) and (d) indicates 1 cm.

same as recipe (1), but the size of each dough ball was proportional to the size of the population in each town (see Fig. 6e); (3) Each FS was formed from a cubic column of agar gel containing oat flakes. The surface area of the column was proportional to the population of town, since the amount of protoplasm accumulating on the FS is proportional to the surface area, if the concentration of ground oat flakes is equal (see Figs. 6a)[12].

Two additional configurations were included to represent the topographic constraints that would normally apply when considering construction of a rail line in the real world. First, a physical obstacle of columnar blocks of plain agar were introduced to correspond to the four major mountain ranges in Hokkaido (see Fig. 7a). Second, the experimental arena was illuminated with a structured mask, as *Physarum* is known to avoid high intensity light [14]. A light intensity of 45000 lx was used to match the position of the four major mountain ranges in the actual Hokkaido landscape (see Fig. 7c).

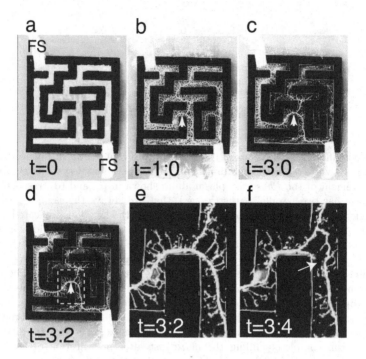

Fig. 3. Solving the maze of topological type a in Fig 1. (a) Top view of the experimental results just after the presentation of the two FSs; (b) at an intermediate state of tube development along the main corridor; and (c, d) at the final state with a single connection through the shortest path. The connection path was stable over a period of time, once the final configuration was reached. In this maze design, there were two possible connection paths between FSs that branched at the point indicated by a white arrow. Interestingly, although the thickest tube was aligned with the shortest path, it did shift a little and then snapped later, as indicated by the white arrow (e,f).

3   Results

3.1   Solving a Maze with the Topological Type in Fig. 1a.

Figure 3a shows results in a maze with the same topology as used previously, but set out differently. Just after the presentation of FSs (Fig 3a), the *Physarum* plasmodium began to gather at the FSs and formed some thick veins along the main corridors in the maze (Fig 3b). Eventually, a single thick vein remained that traced the shortest path between the FSs, whilst the other thick veins disappeared (Fig 3c). This results confirmed that the creature can find the shortest path in the maze, given a choice of two alternative routes, irrespective of the actual physical layout of the maze.

To date, most observations suggest that once a tube path is established, the path does not alter its position. However, in this experiment the tube path was able to shift slightly, as indicated by the white arrow in the time course from Fig 3b to 3d.

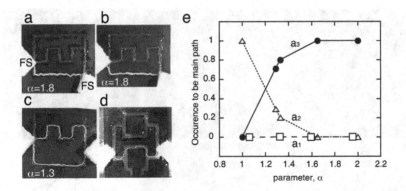

Fig. 4. Statistical occurrence for each path to be main connection between FSs. (a-d) Pictures of the network shape in the maze of Fig. 2a type (a-c) and Fig. 2d type (d). (a) When $\alpha = 1.8$, the shortest path a_3 remained between the two FSs. (b) In some cases in $\alpha = 1.8$, the longer path a_2 was left to connect the FSs but the thickness of a_2 was thinner than that of the main path a_3. We classified the connection through a_3 as main and a_2 as secondary. The path a_2 was preferred to the path a_1 although the length of the two was the same. (c) In some cases in $\alpha = 1.3$, the connections through two paths of a_2 and a_3 were similar, so both paths were classified as main, although the length of a_2 was longer. This showed the typical preference of a_2. (e) Statistical preference to be main path as a function of the parameter $\alpha = \frac{\alpha_2}{\alpha_3} = \frac{\alpha_1}{\alpha_3}$.

The thick tube snapped later, as shown in the white arrow in Fig 3f, and the alternative adjacent thin tube instantly became thicker to compensate for the increased the flow. The observation of snapping strongly suggests that tension forces can act along the tube and may cause re-positioning under some circumstances.

3.2   Solving a Maze with Topological Type in Fig. 1c

With the introduction of additional parallel pathways, including equidistant routes, the potential range of solutions increases. Figure 4 illustrates results of maze-solving for parallel mazes of the type shown in Fig. 1c and Fig. 2a. The path a_3 was always selected when the length was sufficiently shorter ($\alpha > 1.3$) than that of the others, where $\alpha = \frac{\alpha_2}{\alpha_3} = \frac{\alpha_1}{\alpha_3}$. However, path a_2 was always selected when all three paths had the same nominal length. Interestingly, path a_2 was also preferred around 20% of the time when $\alpha = 1.3$, even though it was longer than α_3, and identical to α_2 (Fig. 4c). Figure 4e shows the probability for each route to be selected as the main path as α varies. Moreover, a thin connection through a_2 was sometimes observed even when $\alpha=1.8$, even though the length was much longer (Fig. 4b). Such thin connections were regarded as secondary rather than main or primary. Another interesting observation was that a_1 was not preferred at all although the length of a_1 and a_2 was the same. The results described above implied that additional factors than length alone are involved in path selection.

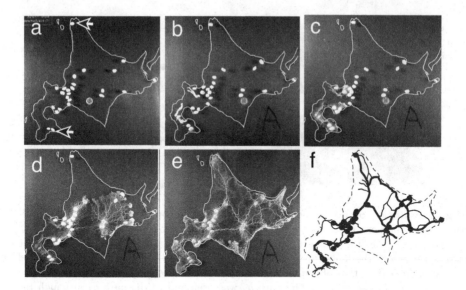

Fig. 5. Time series of the tube-network made by *Physarum* when challenged with 24 equal sized FSs set out in the geographical pattern of the 24 major towns in Hokkaido. (a) Each FS was formed from a dough of oat flakes. The white arrows indicate the two most remote FSs and the distance between these two FSs was 16 cm. Hokkaido is an island surrounded by the sea, so an acetate sheet was cut out to follow the coastline and constrain the organism to fill the land area. (b) A few pieces of plasmodium were initially inoculated in the low left part of Hokkaido. (c, d) Intermediate state of growing network and (e) the final shape of network. (f) Shows the graph representation of network shape obtained from (e). Thick and thin solid lines indicate thick and thin tubes, respectively. The thick-lines represent the main skeleton of the network.

3.3   Transport Network in Hokkaido

Figure 5 shows that *Physarum* connected all FSs into a network throughout the simulated island of Hokkaido. Initially, a few portions of *Physarum* plasmodium were put in the lower left part of the arena (Fig. 5b). The organism started to extend out (Fig. 5c) and fused to become a single organism (d). Finally, the *Physarum* reached all the FSs and connected them with a main skeleton of thick tubes, with a few thinner tubes persisting for a considerable time (e). The network architecture was extracted, after thick tubes and thinner tubes were identified. Figure 5f shows the extracted network, in which thick and thin solid lines indicate the thick and thin tubes, respectively.

Figure 6 shows the tube network connecting FSs with different sizes that scaled in proportion to the population size of each town. A representative typical outcome is shown in Fig. 6c using cubic column of agar gel as FSs and the corresponding network in fig. 6d. In Fig. 6e, the difference in population size was expressed by the size of dough, with the corresponding network shown in Fig. 6f.

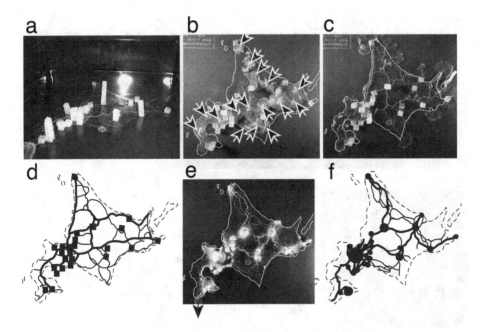

Fig. 6. Time series of the tube-network made by *Physarum* when presented with different sized FSs. (a) A bird's-eye view of the experimental setup in which each FS was cubic column of agar with constant width, but varying height. In general, the surface area of the block scaled with population size, except for the city of Sapporo which was too high to stand. Thus in this case a smaller block but with more oat powder was used. (b) Intermediate stage of growing network. The arrows indicate the locations of organism initially inoculated. (c) The final network that connected all FSs. (d) The network shape extracted from (c). (e) The final network formed when different diameters of oat dough was used as FSs and (f) the corresponding network shape.

Although the network shape was different in each experimental setup, the functional behaviour measured as the total length of the network and its fault-tolerance were similar. The fault-tolerance was defined here as the tolerance of the global connectivity of all FSs to random, accidental disconnection of tube as described previously [8, 11]). These two measures should be traded-off against each other to give a reasonably robust, but not too expensive network. When the size of the FS was varied to reflect population numbers in each location, the skeleton structure remained broadly similar, and still provided a good trade-off between total length and fault-tolerance (Fig. 8).

Figure 7 shows the effects of including mountainous regions on network formation across Hokkaido. Different sized columnar FSs were used as described in Fig. 6, and the mountains were represented by larger blocks positioned to match the four mountain ranges in Hokkaido to prevent the organism exploring these regions (see Fig. 7a). The network structure extracted from Fig. 7b is shown in Fig. 7c. In another experiment, the impact of the mountainous regions was

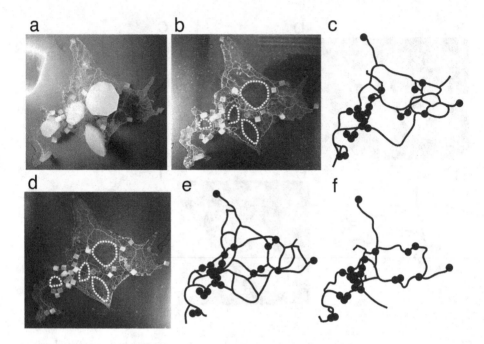

Fig. 7. Effect of including geographical mountainous features on the network formed by *Physarum*. (a) Top view of the experimental setup. Different sized columnar FSs were used, as described in Fig. 6, but with the inclusion of additional non-nutrient blocks of agar corresponding to the four mountain ranges in Hokkaido. (b) The network shape formed around the obstacles, where the thick white dashed lines indicate the locations of the obstacles that were removed after the experiment. (c) Network shape obtained from (b). (d) Network development under patterned illumination (45000 lux) that mapped onto the four mountain ranges (indicated by the thick white dashed lines). The light illumination discouraged growth through the mountainous areas. (e) Extracted network formed in (d). (f) Network layout of the actual railway lines in Hokkaido region in 2010.

simulated using light illumination which *Physarum* tends to avoid. A typical network shape is shown in Fig.7d-e. Figure 7f shows the actual railway network in Hokkaido region in 2010, for comparison with the *Physarum* networks.

4   Path-Finding and Multi-functionally Networking Derived by Morphogenetic Dynamics of Vein

We proposed a simple mathematical model for network formation in the Hokkaido arena. The basic framework of mathematical modeling was extended from the previous model[11] to include the impact of geographical variations of altitude. In brief, the *Physarum* network is represented as a network of water pipes. Fig. 9a show a schematic illustration of model setup with a series of rigid cylindrical pipes connected at junctions, i and j. The pipe is characterized by three physical quantities: length L_{ij}, radius r_{ij}, and flow rate Q_{ij}. Q_{ij} is given by Poisueille flow approximation $Q_{ij} = \frac{\pi r_{ij}^4}{8\eta} \frac{(P_i - P_j)}{L_{ij}}$, where η is viscosity of fluid, and P_i is pressure at the joint i.

At each iteration of the simulation, two joints are selected that correspond to FSs. A constant current Q_0 flows into one of the two and flows out of the other. We can calculate the flow, Q_{ij}, through every pipe, as Q_0 and all L_{ij} and r_{ij} are given, assuming flow follows Kirchoff's law of fluid flow, namely that the sum of currents flowing into each joint is equal to the sum of currents flowing out of that joint.

Next we included the effect of changing vein diameter to simulate the strengthening or disappearance of tubes. The mechanistic basis controlling the dynamics of vein morphogenesis is not well defined, but recent experiments show a strong correlation between the rate of protoplasmic shuttle streaming through the vein itself and its subsequent diameter [15, 16]. Thus veins become thicker when the protoplasmic flow is large enough, but otherwise tend to thin. In essence, each vein adapts locally to the flow through the vein itself that can be described using equations for vein growth [6].

We prefer to use tube conductivity $D_{ij} = \frac{\pi r_{ij}^4}{8\eta}$ rather than just tube radius r_{ij}. Dynamic changes in conductivity follow the balance of two antagonistic processes,

$$\frac{dD_{ij}}{dt} = f(|Q_{ij}|) - \alpha D_{ij},$$

where $f(|Q_{ij}|)$ is the thickening process that depends on the protoplasmic flow and $-\alpha D_{ij}$ is the thinning process with first order reaction kinetics. The function $f(|Q_{ij}|)$ is a monotonically increasing function of the absolute value of Q_{ij}. In this case, we assume $f(|Q_{ij}|) = \tanh(|Q_{ij}| - 1) - \tanh(-1)$. The thinning process comes from the experimental evidence that any local part of organism tends to reduce the intervening network as it accumulates on the FSs.

We now consider the coefficient of the thinning process, α, which can be tuned to reflect geographical features, such as the mountain ranges, lakes and the sea. α is proportional to altitude and scales from $[1 \leq \alpha \leq 3]$ over the range 0 to 2000m. α is also set to maximum over seas, lakes and rivers.

We would like to stress the point that the tube dynamics are controlled locally, with each vein changing according to its own flow and conductivity alone. Nevertheless, the constant Q_0 controls the total flow through the network and provides some indirect coupling across the system. The magnitude of Q_0 plays an important role in the various interactions between the elementary components of these types of models (veins in this case), and is particularly influential on the number of tubes that persist in the simulation. Thus the final network architecture produced by the model is highly dependent on Q_0 and the form of f, as discussed previously [6, 11, 17].

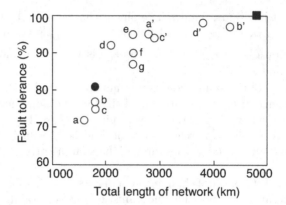

Fig. 8. Evaluation of network performance. Open circles represent experimental data points, with letters representing individual experiments. (a, b) are data from experiments without mountainous regions and equal-sized FSs as shown in Fig. 5; (c, d) include scaling for population size according to Fig. 6a; (e, f) include physical mountainous regions shown in Fig. 7a, whilst (g) includes illumination as shown in Fig. 7d. The 'prime' symbol indicates a data for network that includes the thinner connections, while letters with no prime indicate the skeleton structure of the network with thicker connections. The closed circle and square are results for the real rail and road network, respectively, in Hokkaido region.

5   Model Simulation

Figure 9b-e shows the results of the simulation. The network connects all of the FSs, however, depending on the parameter Q_0, the final shape of network is different, most notably, the total length of network is longer with more parallel paths as Q_0 increases. Nevertheless, the simulated network obtained by the model overlaps the real rail transportation network shown in Fig 7e-g, for relatively low values of Q_0. For comparison with the road network, which has a wide range of different categories of road from major highways to minor side roads, roads were assigned a priority based on their level in hierarchy, and added until total length of lines reached the total length in the simulation. We assume that the highest priority were the highways and descending order of number of national roads (Kokudo) next. The network shape was similar between the model and the real road network with increasing Q_0.

The network tends to avoid the high land, but manages to find routes through the mountain passes acting like a saddle point (indicated by the arrow numbered 1, 2 and 3). Such passes are relatively rare, nevertheless several cities in the south eastern part are connected very effectively with cities in north eastern part of Hokkaido through saddle number 1, and western central part through saddles 2 and 3. These correspond to the real saddle points on the railway and main roads of Bihoro-toge, Karikachi-toge and Nissho-toge, respectively, in Hokkaido.

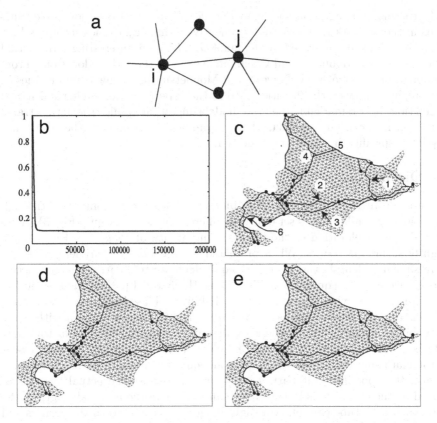

Fig. 9. Model simulation for development of the transportation network between major cities in the Hokkaido region. (a) Schematic representation of the model representing the *Physarum* plasmodium as network of hard cylindrical water pipes, ij, that connect two joints i and j. (b) Time course of the total length of network, that initially covered the whole of Hokkaido as a fine mesh, normalized by the length of initial conditions of network. The total length decreased as the edges disappeared (became thinner than a threshold values of 10^{-5}). During the initial decrease there was some fluctuations in the total length, but it was essentially monotonic. (c) A typical example of the network shape when the total length of the simulated network had stabilized at t=200000 iterations. The numbered arrows 1-3 indicate the correspondence between the simulated network and actual saddle points through the mountain ranges in Hokkaido at Bihoro-toge, Karikachi-toge and Nissho-toge, respectively. The numbered arrows 4-6 indicate the correspondence with actual major triple junctions and the position of accompanying cities of Nayoro, Monbetsu and Oshamanbe, respectively. (d,e) show two solutions run with the same parameters as (c) but with different stochastic choice of sink locations to illustrate that the result is insensitive to this aspect of the simulation. The final network shape is similar in all three cases, and the total length of network is also similar. $Q_0 = 6.0$, $D_{ij}(t = 0) = 1.0$, α is proportional to the altitude of land and $\alpha = 1$ (0 m) and $\alpha = 3$ (2000 m), time step = 0.01, using the Runge-Kutta method to solve the differential equations. The current source, Q_0, is fixed at Hakodate city (indicated by the black arrow), and the current sink is chosen stochastically at every iteration, with the choice probability proportional to the population of the city.

In the model simulation shown in Fig. 9c, there are also major triple junctions in the network that form independently of the major cities set up as FSs. These junctions are indicated by the number 4, 5 and 6. Interestingly the actual cities of Nayoro, Monbetsu and Oshamanbe exist at these three locations. From a population point of view, Nayoro and Monbetsu are the next major cities in the ranking following the 23 cities used in the experimental systems and model simulation. It is remarkable that the model simulation mimics the actual transportation network, and predicts the geographical locations of other cities not explicitly specified in the initial conditions.

6   Discussion

It was confirmed that *Physarum* was able to find the shortest connection through topologically similar mazes set out with different physical layouts (Fig. 3). However, deeper exploration of the maze-solving ability of *Physarum* led to some slightly surprising results. When presented with a series of parallel paths that were similar in length, the notionally equivalent shortest paths were not necessarily selected with equal probability. Thus, the parallel paths a_1 and a_2 in Fig 2a were selected with very different probabilities, biased strongly towards a_2. In contrast, the parallel paths, b_1, b_2 and b_3 in Fig 2b were chosen with equal probability. Taken together, these results suggest that additional factors may influence path selection for the real organism. These factors are also not present in the mathematical model, which gives an equal weighting to paths of the same length. It is possible that path a_2 is preferred because the actual length traversed by the organism is in fact marginally less than the notional length of the corridor, as the tube can follow a slight diagonal route between the corners and keep to the inner curve around the turns. Thus, path a_2 has 7 turns, whilst there are only two in the paths a_1 and a_3. If this is the case, then *Physarum* is able to discriminate a difference in path length of only a few percent of the total. This hypothesis is supported by the preliminary evidence that no preference was observed between three paths in the maze given in Fig. 2d where the number of turning points is the same. It is also possible that topographic features, such as the sharper edges of the turns, provide some additional thigmotropic stimulus, leading to preferential selection of such routes. More comparison between the experiment and the mathematical model is needed to discriminate between these possibilities and determine whether *Physarum* is indeed always able to find the absolute shortest path.

When allowed to form connections between multiple FSs in an arena without constrained paths set up to mimic Hokkaido, the transport network made by *Physarum* showed similar properties to the real infrastructure networks based on total length and fault tolerance, irrespective of (1) the initial locations of the organism; (2) the relative sizes of the FSs; and (3) the different configurations of the FSs (cubic column with different height but same concentration of nutrient, and dough of oat with different diameter). We infer that the core skeleton structure formed by *Physarum* is robust against variations of experimental methods and conditions, provided the physical layout of the FSs is kept constant.

When looking at the core skeleton structure, we tended to ignore the thinner connections, not least because these tended to disappear over extended time periods. However, these connections may be important for the real *Physarum* as they occupy territory and allow the organism to respond to addition or removal of resources. Thus with the stochastic arrival of a new FS, the thin connections may thicken and become part of core skeleton structure. We expect that the thinner connections can work as potential preparation for unknown variations of external conditions in the future.

The total amount of organism initially present is a major factor that determines the extent of the final network. The volume was not controlled precisely in these experiments, but we estimate the variation was within $\pm 50\%$. Under these conditions, The core skeleton structure was similar, with most of the variation in the thinner connections. To experimentally delete the thinner connections, we sometimes put a large FS out side the experimental arena. For instance, in Fig. 6e, the black arrow indicates the location of the large FS. After addition of such a large FS, the thinner tubes disappeared, leaving only the thick tubes. In general, thinner tubes reacted more rapidly to perturbations in the environment, whilst the core skeleton tended to be well conserved and persistent.

The mathematical modeling for the transport network in Hokkaido reproduces the real rail (core) and road infrastructure networks remarkably well. It is impressive that the model reproduces the actual saddle points of Bihoro-toge, Karikachi-toge and Nissho-toge in Hokkaido, which are the most important routes in the actual network connecting the major cities all over Hokkaido. Another interesting feature is that the model can predict the key triple junctions in the network and the accompanying cities of Nayoro, Monbetsu and Oshamanbe. Overall, these degree of correspondence described above implies that similar underlying mechanisms may be common in the social dynamics of humans and amoebae, based on relatively simple principles of the dynamic interplay between the structure of the networks formed and the transport dynamics that unfold upon them. The elegant solutions found by *Physarum* underscore its utility to study how natural self-organized systems can create functionality.

Lastly, we discuss difference of network shape among the experiments, the model simulations and the real configuration of roads/rails. In the experiments, a network shape was totally different case by case if we exactly quantitate the position of each line of trail according to the fixed coordinate axis of Hokkaido space. What this paper showed was that the network shapes were similar even in such variety when we viewed from the evaluation measures of balance between the total length and fault tolerance.

Another point of similarity was that case-by-case fluctuations of passage line coming across the mountain range was relatively smaller in repeats of the experiment, the simulation and the real situations. This means that the constraints of landscape like mountains plays a key role in all of three systems.

We noticed differences of network shape one-by-one. For instance, in the real situations, traffic network covers well around the southern part of Hokkaido including the cities of Hakodate, Esashi, Kaminokuni, Matsumae, but it did not appear both in the model and the experiments. This may be because we ignored the historical process of development in Hokkaido. In fact, these cities were built at the first stage and frontier wave of development propagated from there in some sense.

We pointed out the other factor to be considered although we neglected in this paper. Hokkaido is not isolated in the ocean but have been connected to neighbor islands by huge traffic of matters and people through ship transport. North-east end (Nemuro, Kushiro) and North end (Wakkanai) are such examples and real traffic network is denser than that in the *Physarum* experiment and the model simulation.

In the future, we need a new measure that can characterize the functionality and the physical nature of transport network. Similarity and differences may be much more clarified in the real network and *Physarum* network. Many measures are already proposed in graph theory and so-called network science and, especially, a measure from weighted graph theory, which is not yet developed enough, is to be involved.

Acknowledgements. This research was supported by JSPS KAKENHI 20300105 and by Strategic Japanese-Swedish Research Cooperative Program, Japan Science and Technology Agency (JST).

References

1. Nakagaki, T.: Ph. D. thesis in Nagoya University, Japan (1997),
   http://eprints.lib.hokudai.ac.jp/dspace/handle/
   2115/34739?locale=en&lang=en
2. Nakagaki, T., Yamada, H., Tóth, Á.: Maze-solving by an amoeboid organism. Nature 407, 470 (2000)
3. Nakagaki, T., Yamada, H., Tóth, Á.: Path finding by tube morphogenesis in an amoeboid organism. Biophys. Chem. 92, 47–52 (2001)
4. Nakagaki, T.: Smart behavior of true slime mold in labyrinth. Res. Microbiol. 152, 767–770 (2001)
5. Tero, A., Kobayashi, R., Nakagaki, T.: *Physarum* solver -A biologically inspired method for road-network navigation-. Physica A363, 115 (2006)
6. Tero, A., Kobayashi, R., Nakagaki, T.: Mathematical model for adaptive transport network in path finding by true slime mold. J. Theor. Biol. 244, 553–564 (2007)
7. Nakagaki, T., Yamada, H., Hara, M.: Smart network solutions in an amoeboid organism. Biophys. Chem. 107, 1–5 (2004)
8. Nakagaki, T., Kobayashi, R., Ueda, T., Nishiura, Y.: Obtaining multiple separate food sources: behavioral intelligence in the *Physarum* plasmodium. Proc. R. Soc. Lond. B 271, 2305–2310 (2004)

9. Tero, A., Yumiki, K., Kobayashi, R., Saigusa, T., Nakagaki, T.: Flow-network adaptation in Physarum amoebae. Theory in Biosciences 127, 89–94 (2008)
10. Tero, A., Nakagaki, T., Toyabe, K., Yumiki, K., Kobayashi, R.: A method inspired by Physarum for solving the Steiner problem. International Journal of Unconventional Computing 6, 109–123 (2010)
11. Tero, A., Takagi, T., Saigusa, T., Ito, K., Bebber, D.P., Fricker, M.D., Yumiki, Y., Kobayashi, R., Nakagaki, T.: Rules for biologically-inspired adaptive network design. Science 327, 439–442 (2010)
12. Nakagaki, T., Saigusa, T., Tero, A., Kobayashi, R.: Effects of food amount on path selection in transport network of an amoeboid organism. Topological Aspects of Critical Systems and Networks, 94–100 (2007)
13. Watanabe, S., Tero, A., Takamatsu, A., Nakagaki, T.: Traffic optimization in railroad networks using an algorithm mimicking an amoeba-like organism. *Physarum* Plasmodium, Biosystems 105, 225–232 (2011)
14. Nakagaki, T., Iima, M., Ueda, T., Nishiura, Y., Saigusa, T., Tero, A., Kobayashi, R., Showalter, K.: Minimum-risk path finding by an adaptive amoebal network. Phys. Rev. Lett. 99, 068104 (2007)
15. Nakagaki, T., Yamada, H., Ueda, T.: Interaction between cell shape and contraction pattern. Biophys. Chem. 84, 195–204 (2000)
16. Nakagaki, T., Guy, R.: Intelligent behaviors of amoeboid movement based on complex dynamics of soft matter. Soft Matter 4, 1–12 (2008)
17. Nakagaki, T., Tero, A., Kobayashi, R., Onishi, I., Miyaji, T.: Computational ability of cells based on cell dynamics and adaptability. New Generation Computing 27, 57–81 (2008)

Aggregate "Calculation" in Economic Phenomena: Distributions and Fluctuations

Yoshi Fujiwara

Graduate School of Simulation Studies, University of Hyogo, Kobe 650-0047, Japan
`yoshi.fujiwara@gmail.com`

Abstract. I review recent studies on distributions and fluctuations for personal-income and firm-size in real-economy by using recently available large-data. Specifically explained are Pareto-Zipf laws, Gibrat's law and detailed-balance, and the fact that they are mutually related in a simple way. These patterns and shapes are not "laws" in physics, but can break down in abnormal situations such as bubble-collapse. These findings provide an important foundation of phenomenology for real-economy. The expression "aggregate calculation" nicely fits into the paradigm of this workshop.

Keywords: Pareto-Zipf laws, Gibrat's law, detailed-balance, income distributions, firm-size distributions, growth and fluctuation.

1 Introduction

Phenomenology is crucial for our understanding of natural and human systems. Thermodynamics in physics is a beautiful example of phenomenology on macroscopic systems such as gas, water, light, all materials around us. It took more than a century to establish the phenomenological laws of thermodynamics for systems in non-equilibrium as well as in equilibrium, which later played a crucial role for constructing statistical mechanics based on microscopic principles for equilibrium systems. The role is crucial — statistical mechanics does not give a microscopic foundation of thermodynamics; rather, statistical mechanics is founded on the consistency and compatibility with thermodynamics[1].

Economic phenomena have phenomenological laws. Even if the economy consistes of millions of individuals, firms, banks and other agents at microscopic levels, the aggregate behaviors quite often display patterns and shapes at macroscopic levels. The presence of such patterns and shapes illuminate the validity of a certain phenomenology, although they are not be "laws" in physics. In social science, the emergence of macroscopic patterns from microscopic levels have been lucidly illustrated in many papers and books, for example, "Micromotives and

---

[1] It took me a long time after learning statistical mechanics in university class to recognize the important role of thermodynamics. Thanks to recent textbooks by Professors Yoshi Oono and Hal Tasaki.

Y. Suzuki and T. Nakagaki (Eds.): WSH 2011 and IWNC 2012, PICT 6, pp. 30–38, 2013.

Macrobehavior" by Thomas Schelling. Recently, thanks to abundant electronic data of economic and social systems, researchers able to establish phenomenological findings in more quantitative ways. A good example is high-frequency data in financial markets which have been studies to a great extent. I shall focus on real-economy in this paper.

Specifically I shall show that

- distributions and fluctuations of personal income and firm-size have interesting phenomenological patterns and shapes;
- they are Pareto-Zipf laws, Gibrat's law and detailed-balance, which are mutually related in a simple way;
- but the patterns and shapes break down under abnormal situations in economy,

by using large-data of personal income and firm-size.

These patterns and shapes of distributions and fluctuations are quite robust in the sense that they can be observed for different choice of variables, for various industrial sectors and countries, even if they can break down. This phenomenological fact implies that the finding is linked to the dynamics of aggregate behavior of individuals and firms. Each individual or firm may be in growth, shrink and in birth-death (entry-exit); but, stochastically, the aggregated system is relatively stable unless it is subject to large shocks. In my opinion, it looks as if the system were "calculating" in a self-organized way. Honestly I do not know how to define calculation, say in biological system or in social system, but I think that the word "aggregate and stochastic calculation" nicely expresses the phenomenology I shall explain in what follows.

The figures in this paper were reproduced from the publication [1, pp.24–33]. Interested readers are also suggested to read the book [2] further.

2   Distributions

The distributions for personal income or firm-size are strikingly different from those for human heights, for example. One has no chance to observe persons 1km tall nor 1mm small. The universe of personal income or firm-size is dominated by giants and dwarfs — *a few giants and many dwarfs* — in the terminology of economics. Two examples of such distributions are shown in Fig. 1 and Fig. 2.

Let x be personal income or firm-size. Cumulative probability distribution $P_>(x)$ is the probability that a given individual or firm has income or size equal to, or greater than x. It is obvious from Fig. 1 and Fig. 2 that for large x one has a power-law:

$$P_>(x) \propto x^{-\mu} , \tag{1}$$

where μ is a constant, often called Pareto index. This phenomenon, now known as Pareto law, has been widely observed. μ is typically around 2 for personal income and around 1 for firm size distribution. The latter case is often referred to as Zipf law. We call the distribution (1) Pareto-Zipf law in this paper.

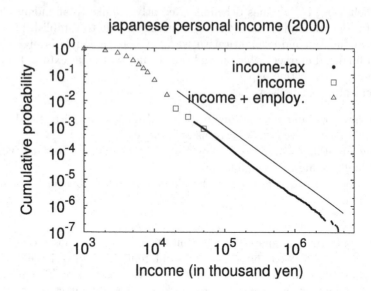

Fig. 1. Cumulative probability distribution of Japanese personal income in the year 2000. The line is simply a guide for eyes with $\mu = 1.96$ in (1). Note that the dots are income tax data of about 80,000 taxpayers. Reproduced from [1].

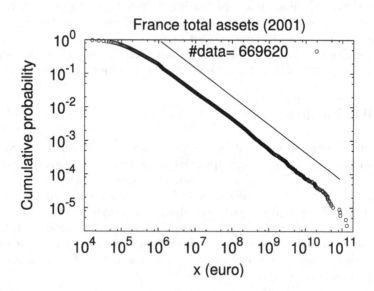

Fig. 2. Cumulative probability distribution of firm size (total-assets) in France in the year 2001. Data consist of 669,620 firms, which are exhaustive in the sense that firms exceeding a threshold are all listed. The line corresponds to $\mu = 0.84$ in (1). Reproduced from [1].

Note that $P_>(x)$ is a cumulative probability distribution function (cumulative PDF). The probability distribution (density) function (PDF), $P(x)$, is related to $P_>(x)$ as

$$P(x) = -\frac{dP_>(x)}{dx} , \qquad (2)$$

by definition.

Understanding the origin of the law has importance in economics because of its linkage with consumption, business cycles, and other macro-economic activities. Also note that even if the range for which (1) is valid is a few percent in the upper tail of the distribution, it is often observed that such a small fraction of individuals (firms) occupies a large amount of total sum of income (size). Small idiosyncratic shock can make a considerable macro-economic impact.

3   Fluctuations

Pareto's power-law is commonly found in many phenomena observed within natural science, as well as the social and economic sciences. Most researchers in these fields would accept that there is no single mechanism giving a universal explanation for all these phenomena, but rather that different mechanisms would be found for each of a variety of classes of phenomena. Mechanisms explaining the origins of power-laws in natural and social sciences have been uncovered for more than a dozen classes of phenomena. See references given in [2, Chapter 3].

Personal income and firm-size obviously change over time: last year's sales are not equal to this year's. Thus we can see that observations on the distribution of company size are an instantaneous snapshot of the state of a collection of companies, each of which is individually subject to fluctuations.

Let us denote by x_1 and x_2 a firm's sizes at time t_1 and a succeeding time t_2 respectively (or a person's incomes). To examine the temporal change, we define growth-rate by

$$R = \frac{x_2}{x_1} . \qquad (3)$$

The variable R represents the ratio at which a company increases or decreases its size, such as s ales and total-assets, in the time interval. In addition, we will define the logarithmic growth-rate as

$$r = \log_{10} R = \log_{10} x_2 - \log_{10} x_1 , \qquad (4)$$

where $\log_{10}$ is a logarithm with base 10. Thus $r = 1$ and $r = -1$ correspond to $R = 10$ and $R = 0.1$ respectively.

By identifying each high-income earner and large-size firm, one can directly observe the temporal change from x_1 to x_2 at one year and the next year, say. The examples are shown in the scatter plots of Fig. 3 and Fig. 4.

The scatter plot represent the joint distribution $P_{12}(x_1, x_2)$ for the pair of variables x_1 and x_2. The plots in Fig. 3 and Fig. 4 are consistent with what is

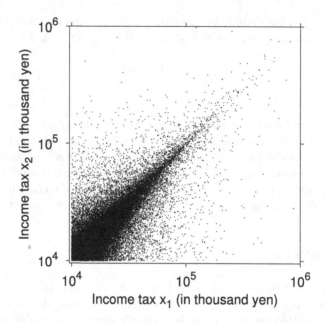

Fig. 3. Scatter-plot of all individuals whose income tax exceeds 10 million yen in both 1997 and 1998. These points (52,902) were identified from the complete list of high-income taxpayers in 1997 and 1998, with income taxes x_1 and x_2 in each year. Reproduced from [1].

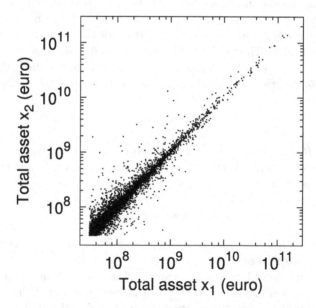

Fig. 4. Scatter-plot of the firms in France whose firm size (total-assets) exceeds a certain threshold in both 2000 and 2001

called *detailed-balance* in the sense that the joint distribution is invariant under the exchange of values x_1 and x_2, i.e.

$$P_{12}(x_1, x_2) = P_{12}(x_2, x_1) \ . \tag{5}$$

One can actually perform a statistical test for the symmetry in the two arguments of $P_{12}(x_1, x_2)$ by two-dimensional Kolmogorov-Smirnov test. Detailed-balance means that the empirical probability for an individual to change its income from a value to another is statistically the same as that for its reverse process in the ensemble.

4   Distributions of Growth-Rates

Finally we shall take a look at the distributions of growth, R or equivalently r. We examine the PDF for the growth rate $P(r|x_1)$ on the condition that the income x_1 in the initial year is fixed. By doing this, we can see whether the growth of personal income or firm-size depends on the starting income or size of x_1. In other words, do giants grow at faster or slower pace than dwarfs do?

The distributions for such growth rates are shown in Fig. 5 (a) and Fig. 5 (b). Note that the different curves collapse into a single curve of distribution in each of the figures. This means that the distribution for growth rate r is statistically independent of the value of x_1. This is known as *law of proportionate effect* or *Gibrat's law* [3].

This can be stated mathematically as

$$P(r|x_1) = \text{a function of } r \text{ only} \equiv Q(r) \ , \tag{6}$$

where $Q(r)$ is a function of r, which has different functional forms for different variables.

The probability distribution for the growth rate, such as the one observed in Fig. 5, contains information of dynamics. One can notice that it has a skewed and heavy-tailed shape with a peak at $R = 1$. How is such a functional form consistent with the detailed-balance shown in Fig. 3? And how these phenomenological facts are consistent with Pareto's law in Fig. 1? Answers to these questions are given in the next section.

5   Relations among the Laws

To summarize the empirical findings, one has the Pareto's law in (1), the detailed-balance in (5), and the Gibrat's law in (6). We shall show that Gibrat's law and the detailed balance lead to Pareto's law.

Since the pair of variables (x_1, x_2) and that of (x_1, R) are related by the change of variable, $R = x_2/x_1$, one can easily see that the joint probability distribution $P_{1R}(x_1, R)$ is related to the joint probability distribution $P_{12}(x_1, x_2)$ by

$$P_{12}(x_1, x_2) = \frac{1}{x_1} P_{1R}(x_1, R) \ . \tag{7}$$

(a)

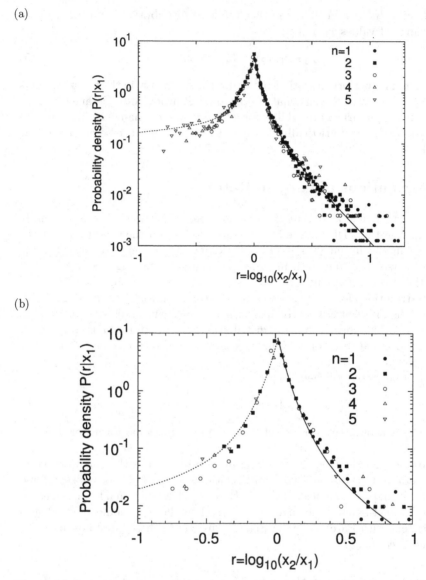

(b)

Fig. 5. (a) Probability density $P(r|x_1)$ of growth rate of individual income in Japan, $r \equiv \log_{10}(x_2/x_1)$, from 1997 to 1998 for income exceeding a threshold. Different bins of initial income-tax with equal size in logarithmic scale were taken to plot probability densities separately for each such bin $n = 1, \cdots, 5$. The solid line in the portion of positive growth ($r > 0$) is an analytic fit. The dashed line ($r < 0$) on the other side is calculated by the relation in (12). Reproduced from [1]. (b) Probability density $P(r|x_1)$ of growth rate of firm-size in France, $r \equiv \log_{10}(x_2/x_1)$, from 2000 to 2001 for size exceeding a threshold. Different bins of initial income-tax with equal size in logarithmic scale were taken to plot probability densities separately for each such bin $n = 1, \cdots, 5$. The solid line in the portion of positive growth ($r > 0$) is an analytic fit. The dashed line ($r < 0$) on the other side is calculated by the relation in (12).

Now, the conditional PDF $P(R|x_1)$ for the growth-rate satisfies

$$P_{1R}(x_1, R) = P(R|x_1)\,P(x_1)\,, \qquad (8)$$

by definition, where $P(x_1)$ is the PDF for the size x_1.

Assume that detailed balance holds, then (7) yields

$$P_{1R}(x_1, R) = \frac{1}{R}P_{1R}(Rx_1, R^{-1})\,, \qquad (9)$$

as readily shown by a simple calculation. Therefore, (9) and (8) lead us to

$$\frac{P(R^{-1}|x_2)}{P(R|x_1)} = R\frac{P(x_1)}{P(x_2)}\,. \qquad (10)$$

Note that this is a consequence from detailed balance alone.

With the additional assumption of Gibrat's law in (6), one can immediately rewrite (10) as

$$\frac{P(x_1)}{P(x_2)} = \frac{1}{R}\frac{Q(R^{-1})}{Q(R)}\,. \qquad (11)$$

Note that the left-hand side of (11) is a function of x_1 and $x_2 = R\,x_1$, while the right-hand side of (11) is a function of R. The equality holds if and only if $P(\cdot)$ is a power-law function, $P(x) \propto x^{-\mu-1}$, where μ is a constant. (For example, expand (11) in terms of R around $R = 1$, and obtain a differential equation that $P(\cdot)$ has to satisfy, which can be solved easily.) By integrating (2), one has the Pareto's law in (1).

Furthermore, by inserting the power-law function $P(x) \propto x^{-\mu-1}$ into (11), one has

$$Q(R) = R^{-\mu-2}Q(R^{-1})\,, \qquad (12)$$

which relates the positive and negative growth rates, $R > 1$ and $R < 1$, through the Pareto index μ. This is a non-trivial consequence of our argument here, and can be checked for its validity in the real data. See Fig. 5 (a) and Fig. 5 (b).

Therefore, the phenomenological properties (A) detailed-balance, (B) Pareto-Zipf law, and (C) Gibrat's law are observed for firm size as well as for personal income.

6   Summary

We have shown the following stylized facts concerning distribution of personal income and firm size, their growth and fluctuations by studying exhaustive lists of high-income persons and firm sizes in Japan and in Europe.

- In power-law regime, detailed-balance and Gibrat's law hold.
- Under the condition of detailed-balance, Gibrat's law implies Pareto's law.
- Growth-rate distribution has a non-trivial relation between its positive and negative growth sides through Pareto index.

The empirical "laws" of Pareto, Gibrat, and detailed-balance are not laws in physics, but patterns and properties of economics. They do not hold in non-power-law regimes, or can break down when the economy is under abnormal conditions.

- Power-law, detailed-balance and Gibrat's law break down according to abrupt change in risky asset market, such as Japanese "bubble" collapse of real estate and stock.
- For firm size in non-power-law regime corresponding to small and middle size firms, Gibrat's law does not hold. Instead, there is a scaling relation of variance in the growth-rates of those firms with respect to firm size, which asymptotically approaches to non-scaling region as firm size comes to power-law regime.

See [1,2] and references therein.

The stylized facts that we described in this paper, however, serve as an established phenomenology, which any models for households and firms activities should satisfy.

Acknowledgment. This work is partially supported in part by the Program for Promoting Methodological Innovation in Humanities and Social Sciences by Cross-Disciplinary Fusing of the Japan Society for the Promotion of Science. The authors would like to thank Hideaki Aoyama, Mauro Gallegati and Corrado Di Gulimi, Yuichi Ikeda, Hiroshi Iyetomi, Wataru Souma for collaborations.

References

1. Fujiwara, Y.: Pareto-Zipf, Gibrat's laws, detailed-balance and their breakdown. In: Chatterjee, A., Yarlagadda, S., Chakrabarti, B.K. (eds.) Econophysics of Wealth Distributions. Springer, Italia (2005)
2. Aoyama, H., Fujiwara, Y., Ikeda, Y., Iyetomi, H., Souma, W.: Econophysics and Companies: Statistical Life and Death in Complex Business Networks. Cambridge University Press (2010)
3. Sutton, J.: Gibrat's legacy. J. Economic Literature 35, 40–59 (1997)

Towards Co-evolution of Information, Life and Artificial Life

Masami Hagiya and Ibuki Kawamata

Department of Computer Science,
Graduate School of Information Science and Technology,
The University of Tokyo 7-3-1 Hongo, Bunkyo-ku, Tokyo, 113-8656, Japan
{hagiya,ibuki}@is.s.u-tokyo.ac.jp

Abstract. We will begin with a simplified view of systems biology and synthetic biology. Systems biology extracts information from life, while synthetic biology converts information to reality. This cycle allows the co-evolution of life and information, and accelerates the evolution of both. Additionally, the field of molecular robotics has recently emerged. This field is attempting to implement artificial life using biological molecules. We foresee that molecular robots will interface information and life, and the distinction among information, life and artificial life will eventually become a blur. Once molecular robots gain the ability to evolve, then co-evolution of the three will lead to a new stage of intelligence.

1   Introduction

We are currently in an era in which life and information can evolve together. Among the fields of research that involve both information technology and biology, bioinformatics and systems biology analyze "big data" obtained by cutting-edge bio-imaging technologies [8], and construct information (models and databases) on life on top of computers and networks, recently described as the "cloud". Conversely, synthetic biology, another emerging field, attempts to construct reality from models designed by humans [7]. If systems biology is regarded as reverse engineering of life, then synthetic biology advocates forward engineering of life. These research fields thus form a cycle of life and information, as described in Figure 1.

The implication of this cycle is profound. Life has evolved for billions of years and adapted to the environments of the earth. This is a type of "optimization", which consists of extremely complex processes. In addition, components at a lower level of hierarchy of life self-organize and express behaviors at an advanced level; this phenomenon is called "emergence". Based on this cycle, evolution and self-organization of life can be replaced with evolution and self-organization of information. Furthermore, co-evolution of life and information is possible. This situation is depicted in Figure 2. In addition to the original route, there exists another route for evolution and self-organization in life, where life is first converted to information, which evolves and self-organizes, and is then converted back to life. It is important to note that many existing methods for evolution and

Y. Suzuki and T. Nakagaki (Eds.): WSH 2011 and IWNC 2012, PICT 6, pp. 39–48, 2013.

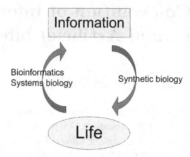

Fig. 1. Cycle around life and information

self-organization of information, such as evolutionary computation, are borrowed or inspired by evolution and the self-organization of life. The two routes may act in parallel to create new kinds of life. As a result, the evolution of life is accelerated compared with the situation where the two routes are independent, because a single step of evolution in one route may trigger a few steps of evolution in the other and result in exponential growth in the sense of singularity.

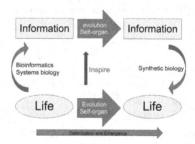

Fig. 2. CyCo-evolution of life and information

In this position paper, we will first describe the spectrum of wet artificial life, including synthetic biology and molecular robotics, in Section 2. Research fields in this spectrum attempt to construct complex artificial systems from molecules; in particular, biomolecules such as DNA and protein. According to synthetic biology, components from life can be borrowed and modified. Some research fields, such as molecular robotics, attempt to construct life-like systems from basic components. However, these research fields share a single goal. In Section 3, the evolution and self-organization of information are briefly sketched. We classify them into two types: those by artificial intelligence and those by human intelligence, and put them in the context of the cloud, where various types of intelligence cooperate. We then envision the future from the perspective of molecular robotics in Section 4. We foresee that molecular robots will interface information and life, and the distinction among information, life and artificial life will eventually blur. After molecular robots gain the ability to evolve, co-evolution of the three will lead to a new stage of intelligence.

2 Spectrum of Wet Artificial Life

Channon et al. summarized and classified the approaches to synthetic biology, as described in Figure 3, which is simplified from the original [2]. On the left of the figure, various levels in the biological hierarchy from simpler to more complex are shown, with examples. The x-axis denotes the "unnaturalness" of synthesized systems. This figure clearly shows the direction towards constructing more complex and unnatural systems starting from natural and simple ones. Needless to say, it covers more than synthetic biology in the narrow sense. For example, they include organic chemistry and supra-molecular chemistry. In particular, the emerging field of molecular robotics is introduced in the next section. The ultimate goal of these research fields is "encapsulated complex systems", as shown in Figure 3. The two arrows represent synthetic biology (in the narrow sense) and molecular robotics, respectively.

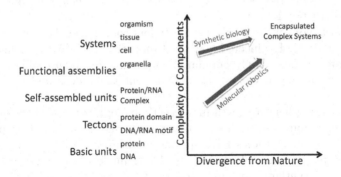

Fig. 3. Spectrum of synthetic biology

2.1 Molecular Robotics

Molecular robotics is an emerging research field whose aim is to construct autonomous systems from molecules. Autonomous systems suggest those composed of sensors, computers (information processing units) and actuators. Signals from the external environment are recognized by the sensors and transformed into internal signals fed to the information processing units, which perform computations and make judgments that order the actuator to perform its functions. In addition to these three components, an autonomous system should include encapsulating structures, as the phrase "encapsulated complex systems" in Figure 3 suggests. Energy sources that drive the three components are also necessary.

Presently, molecular robotics utilizes mainly DNA molecules [1], which is why the field is also described as DNA robotics. There are many reasons for the use of DNA: DNA can form various structures in terms of hydrogen bonds between complementary sequences [10], which can be rationally designed and chemically synthesized; and DNA also allows various chemical modifications. Some sequences have chemical activity and can be used as sensors or actuators [3]. Therefore, molecular robots can be constructed using only DNA molecules. Such DNA robots may be inefficient for concrete applications, but are convenient for prototyping. Currently, an increasing number of chemists are joining the field, so additional types of molecules will likely be used in future applications.

2.2 Synthetic Biology

Cells possess all the necessary functions for autonomous systems. For example, membrane receptors are sensors. Information processing inside a cell is completed by genetic circuits, signal transduction pathways, etc. There are also several types of actuators within a cell. Protein synthesis per se is a type of actuation. Movement of a cell, such as chemotaxis, is also a type of actuation. In addition to these functions, cells themselves can also reproduce. Therefore, why not make robots by re-engineering cells? This idea has led to the emerging research field known as synthetic biology [7].

In synthetic biology, the international student competition known as the International Genetically Engineered Machine Competition (iGEM), is held annually and is becoming increasingly popular. Many teams of undergraduate students from all over the world congregate and present their ideas and experimental results.

As an emerging field, research in synthetic biology is presently very diverse. For example, various case studies are being conducted, including those presented at iGEM. Some case studies have specific applications, such as drug delivery and bio-fuel production. Efforts are also being made to introduce engineering disciplines into synthetic biology, such as standardization and abstraction of genetic components. Some researchers are interested in the construction of a minimal cell, which contains the minimum set of genes required for survival. Technology for the swapping of genomes in a cell facilitates production of a cell with a completely designed genome. Attempts are also being made to construct a cell from scratch. This is closely related to molecular robotics, and involves the technology required for generation of artificial membranes.

3 Evolution and Self-organization of Information

Evolution and self-organization of information is diverse. Here, they are classified into two types: those by human intelligence and those by artificial intelligence. Both have been amplified or powered by recent developments in information technology. Needless to say, these two types are not separate, but are interrelated. As we will discuss later, they even cooperate in the cloud.

3.1   Amplification of Human Intelligence

Although the phrase Web 2.0 is becoming outdated, collaboration in the cloud is becoming more common, resulting in so-called collective intelligence. While Wikipedia is a typical example, synthetic biology is also a good example of collective intelligence. The organizers of iGEM are constructing a genetic component database (called parts) for use by participants of the competition, who should submit their parts to the database while being allowed to freely use all parts therein. The organizers thus accumulate knowledge for use in future applications, such as medicine, environment preservation and food production.

In general, collaboration in the cloud is becoming increasingly fine-grained and automated. In terms of wet experiments in synthetic biology, pieces of information that were typically kept in laboratory notebooks are now being stored in the cloud in a partially-automated fashion. This is in contrast to traditional biological databases in which results from a series of experiments were stored in conjunction with publications after completion of the related experiments.

We are currently developing a database for wet biological experiments and using it for our own experiments in the fields of synthetic biology and molecular robotics [4]. This database is based on MediaWiki, a free, open-source wiki package originally developed for Wikipedia. A snapshot of a project page is shown in Figure 4. A project page describes a schedule of experimental steps, such as those beginning with plasmid preparation by infusion and ending with plasmid extraction by a miniprep procedure. A schedule of experimental steps is shown as a workflow graph, from which pages keeping information on materials and pages for concrete experiments are linked. We use MediaWiki templates extensively so that the minimal description of a project, such as that in Figure 5, suffices to automatically generate a workflow graph and related material and experimental pages. The database can also be used with a tablet computer, such as an iPad, so that it replaces laboratory notebooks during wet experiments. We are also developing interfaces that control experimental devices to automate specification of parameters and extraction of data.

Using this database, we have conducted wet experiments in synthetic biology. They include almost all basic experimental steps in synthetic biology, including transformation, gel electrophoresis, photospectroscopy, infusion, mutagenesis, sequencing, minipreps, etc. We also performed experiments in which AND gates were implemented in Escherichia coli.

We hope that this database will eventually control robots and lead to full automation of wet experiments. Dry experiments, such as prediction of the energy parameters of DNA strands, can already be conducted using the database, and experimental results are automatically inserted therein. External services, such as DNA synthesis and plasmid sequencing, can also be linked from the database through the web interface.

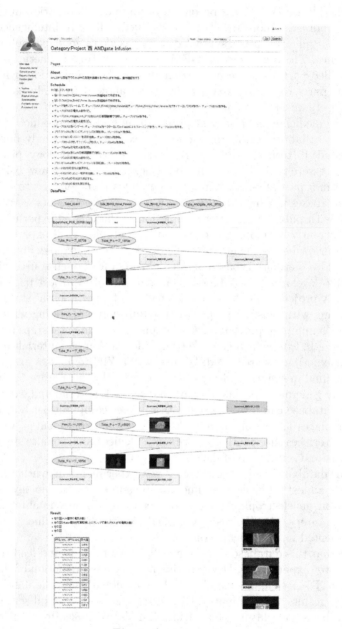

Fig. 4. A project page

Fig. 5. Source code of the project page in Fig 4

3.2   Automatic Design by Artificial Intelligence

By artificial intelligence, we mean automatic design of systems, in general. For example, in the systems biology field, it is common to predict genetic circuits from microarray data. For this purpose, evolutionary computations, such as genetic algorithms and particle swarm optimization, are commonly used. In addition to predictions, they can also be used for the design and synthesis of molecular and cellular systems.

Evolutionary computation, the most typical of the so-called bio-inspired methods, borrows the ability of life to evolve and optimizes its functions with respect to its environment. For example, in genetic algorithms, evolution by mutation, crossover and selection are mimicked in silico to solve various search and optimization problems. Compared with other optimization methods, this is more effective when the search space is large and the evaluation function is complex, and it has been applied to various design problems, including those of DNA robots and artificial genetic circuits. In the latter applications, we can say that methods inspired from life create life!

A concrete example of applying evolutionary computation to molecular robotics can be found in the work by Kawamata et al. [5]. They employed simulated annealing to design some DNA devices including logic gates. A specification of a device is first given and DNA devices are randomly generated at the level of DNA segments consisting of ten to twenty bases. Generated devices are then simulated and evaluated with respect to the specification. Those devices with a good evaluation value are chosen and improved by random mutation.

It seems interesting to combine such in-silico evolution with in-vitro evolution of molecules. For example, in-silico evolution may also introduce artificial molecules having hypothetical functions. If a device containing such an artificial molecule has a high evaluation value, it is worth while to look for a real molecule having the specific function by in-vitro evolution. If new molecules are actually found by in-vitro evolution, they are added to the library of molecules for constructing molecular devices and increase the variety of possible devices.

3.3 Cloud

Cloud computing started with shared computer resources, such as storage, CPU time and networks. As mentioned earlier, resources shared in the cloud are becoming increasingly fine-grained. For example, while entire virtual machines are shared in Amazon EC2, single files are shared in distributed storage systems such as Dropbox. In our database of biological experiments, we are also attempting to store and share in the cloud more fine-grained information regarding each experimental step.

In addition, an increasing number of services are connected to the cloud, including laboratory devices and external services. The cloud is also where artificial intelligence and human intelligence cooperate. We depict this situation in Figure 6.

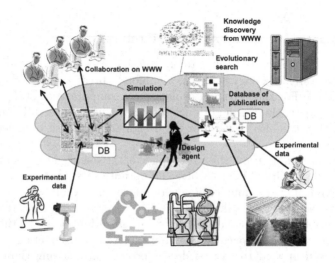

Fig. 6. The cloud

4 Future of Molecular Robots

We have described the vision that under the cloud, life and information evolve together. However, the interface between life and information is somewhat large-scale and is often subjected to human intervention. More direct interfaces between the two are expected in the future.

Murata et al. recently summarized the evolution scenario of molecular robots [9]. They explain the evolution of molecular robots in several stages. In the current (0th) stage, single-molecular robots like DNA spiders have been constructed, but their movement is heavily ruled by Brownian motion. In the first stage, in order to control or utilize the randomness of molecular reactions, compartments such as artificial membranes will be created, leading to amoeba-type molecular robots. In the second stage, slime-type robots, whose size will reach the millimeter scale, will utilize the three-dimensional chemical wave field in a gel for movement. In the third stage, multi-cellular molecular robots will be created, and in the fourth stage and on, hybrid molecular robots that directly interface with electronic computers will emerge.

4.1   Connecting Chemical-Biological and Electronic Systems

Although foreseen in the last stage of the evolution of molecular robots, directly connecting or even merging chemical-biological and electronic systems has already become one of the biggest trends in science and technology, including BMI (brain-machine interface), MEMS (micro-electro-mechanical system), ultrafast DNA sequencing, and bio-imaging. Molecular robots are expected to solve many problems currently faced by human society. We believe that molecular robotics should participate in this trend as early as possible.

Electronic computers are thought to face a limitation in terms of miniaturization. Molecular electronics was once regarded as a key technology in construction of nanometer-scale circuits. However, the problem of how to connect molecular transistors with themselves or silicon circuits remains. Although DNA nanotechnology (DNA origami) was expected to solve this problem, there remain technical obstacles to overcome [6].

In contrast, the miniaturization technology of VLSI has progressed, and currently, the so-called 20nm-process is used for production. Electronic computers now face an energy dissipation problem, which is closely related to computation errors, because a reduction in energy leads to a loss of exactness with each switching. Further miniaturization should also increase error probability.

4.2   Dream

In this era of parallel computing, the fastest supercomputers are composed of a huge number of many-core nodes connected by ultrafast networks. What types of application are run on such supercomputers? Scientific simulation remains one of the most important applications, but more computing resources are used for machine learning, searching and optimization, in which evolutionary computation is also included. These types of computers tolerate errors at some level. In reality, parallel computation often reports errors in both hardware and software.

Therefore, error-tolerant computation, such as neural networks, are likely the key to future computers. Thinking about neural networks, we are tempted to make an analogy with the brain, where computation is conducted by the transmission of electric pulses of neurons, while learning takes place through changes in synaptic connections, which are molecular in nature and very slow compared to computation.

With this analogy in mind, we propose the architecture (or dream) of a molecular-electronic chip. This consists of neurons composed of VLSI circuits that are laid out on one side of the chip. Neurons contain terminals that go pass through to the other side of the chip, which is immersed within a solution in which molecular robots move. One side of the chip is dry while the other is wet. Molecular robots may stick to some terminals and transmit electric current themselves or put in place molecules to connect terminals. Computation is conducted by electronic current along the circuits and connections between terminals, while learning is carried out by molecular robots. Note that this chip also realizes the trend of merging chemical-biological and electronic systems.

We can generalize the above dream and foresee a future in which molecular robots gain the ability to evolve. As they become a part of electronic computers, both the information stored therein and the computers themselves evolve. In addition, molecular robots interface with cells. This means that life and artificial life evolve together. Therefore, co-evolution of information, life and artificial life will become possible, and will lead to a new stage of intelligence, while the borders among the three will eventually disappear.

References

1. Bath, J., Turberfield, A.J.: DNA nanomachines. Nat. Nanotechnol. 2(5), 275–284 (2007)
2. Channon, K., Bromley, E.H.C., Woolfson, D.N.: Synthetic biology through biomolecular design and engineering. Curr. Opin. Struct. Biol. 18(4), 491–498 (2008)
3. Cho, E.J., Lee, J.W., Ellington, A.D.: Applications of aptamers as sensors. Annu. Rev. Anal. Chem. 2, 241–264 (2009)
4. Kawamata, I., Hagiya, M.: (2012), http://hagi.is.s.u-tokyo.ac.jp/syn-biol/
5. Kawamata, I., Tanaka, F., Hagiya, M.: Automatic Design of DNA Logic Gates Based on Kinetic Simulation. In: Deaton, R., Suyama, A. (eds.) DNA 15. LNCS, vol. 5877, pp. 88–96. Springer, Heidelberg (2009)
6. Kershner, R.J., Bozano, L.D., Micheel, C.M., Hung, A.M., Fornof, A.R., Cha, J.N., Rettner, C.T., Bersani, M., Frommer, J., Rothemund, P.W.K., Wallraff, G.M.: Placement and orientation of individual DNA shapes on lithographically patterned surfaces. Nat. Nanotechnol. 4(9), 557–561 (2009)
7. Khalil, A.S., Collins, J.J.: Synthetic biology: applications come of age. Nat. Rev. Genet. 11(5), 367–379 (2010)
8. Kitano, H.: Systems Biology: A Brief Overview. Science 295(5560), 1662–1664 (2002)
9. Murata, S.: Molecular robotics: A new paradigm for artifacts. New Generation Computing 31(1), 27–45 (2013)
10. Seeman, N.C.: Nanomaterials based on DNA. Annu. Rev. Biochem. 79, 65–87 (2010)

Harness the Nature for Computation

Yasuhiro Suzuki

Department of Complex Systems Science, Graduate School of Information Science,
Nagoya University, Furocho Chikusa Nagoya City 464-8601, Japan
ysuzuki@nagoya-u.jp

Abstract. *Natural computing* investigates and models computational techniques inspired by nature and attempts to understand natural phenomena as information processing. In this position paper, we consider harness the nature for computation, from the perspective of natural computing. We investigated facsimile computational models of self-organization in nature, and identified dissipation of information flow as a common mechanism, where intermediate information is produced through interactions and consumed through evoking novel interactions. Based on this mechanism, we propose the concept of a harness: an indirect controlling method for natural systems. We realize this concept through a computational model, and discuss how this concept has already been successfully applied in medical and ecological science.

Keywords: Harness, Dissipation of Information, Self-organization.

1   Introduction

The principle of natural computing lies in understanding nature in terms of computing[10]. In order to understand a natural phenomenon, we mustdefine all related elements and interactions precisely; otherwise,we shall not be able to obtain the algorithm for the phenomenon. However, it isvirtually impossible for humans to amass all the knowledge required for such purposes.

Although computing has been used in activities such as sheep husbandry, it is not possible to know everything about sheep, e.g., the manner in which they communicate. It is known that sheep are flock animals with a natural inclination to follow a leader; therefore, a shepherd or sheepdog can easily control a flock of sheep by intimidation. Shepherds adopt strategic intimidation with a specific objective; we refer to such a strategy as a "harnessing." A harness is a sequence of instructions that can be modified to alter sheep behaviour. Hence, we can design a harness for guiding a flock by changing the order of instructions. Therefore, a harness can be regarded as an algorithm for computation. However, it is different from a conventional algorithm, which requires precise knowledge of all the related elements and interactions. Using a harness, a shepherd can control a flock of sheep without knowing all about the sheep and their interactions.

To design a harness, we must have precise information about the instructions to be used; the construction of a harness is similar to that of a computer algorithm. When the designed harness is applied to natural systems, it may have

Y. Suzuki and T. Nakagaki (Eds.): WSH 2011 and IWNC 2012, PICT 6, pp. 49–70, 2013.

desired as well as undesired effects, thereby altering the system behaviour. Such effects and the resulting changes in behaviour may be unexpected.

Hence, computation using a harness differs from that using an algorithm. The execution of sequential instructions is similar in both cases; however, in the case of a harness, the computation result does not solve the problem directly, but produces effects and side effects in the system and guides the system to solve the problem.

In this paper, we consider the use of a harness for natural computing by investigating an algorithm for spontaneous self-organization in nature, on the basis of an artificial chemistry known as abstract rewriting system on multisets (ARMS).

2   Abstract Rewriting System on Multisets, ARMS

An Abstract Rewriting System on Multisets (ARMS, [14]) is a model of artificial chemistry [1] based on computational algebra (rewriting systems) and physical chemistry (such as Gillespie's method [5]). We computationally characterized the Edge of Chaos [15] by investigating the relationship between ARMS and a cellular automaton. We showed how (computational) living things emerge through chemical evolution and can be evolved using an ARMS, by applying this evolutionary system to solve a simple mathematical problem [16]. Furthermore, we propose a model of evolutionary dynamics for the proto-enzyme through an evolutionary reaction network modeled by an ARMS, where repeated auto-catalytic reaction networks emerge and are catastrophically destroyed [19]. This type of behavior has also been reported using a replicator system [2]. Beyond the field of artificial life, ARMS has also been used in ecology [18], medical science [21] and environment engineering [6] among others.

ARMS is calculated as an expression of expressed of the Chemical Master Equation (CME), a stochastic expression of the Reaction Rate Equation (RRE). Here, we demonstrate that an ARMS can be regarded as a CME and, through continuous approximation, the deterministic RRE, which is denoted by a set of ordinal differential equations that can be obtained from an ARMS [17].

Fundamentally, an ARMS is a construct $\Gamma = (A, w, R)$, where A is an *alphabet*, w is a multiset present in the initial configuration of the system, and R is the set of multiset rewriting rules. Let A be an alphabet (a finite set of abstract symbols). A multiset over A is a mapping $M : A \mapsto \mathbf{N}$, where N is the set of natural numbers; 0, 1, 2,..... For each $a_i \in A$, $M(a_i)$ is the multiplicity of a_i in M. We can also denote $M(a_i)$ as $[a_i]$. We denote by $A^{\#}$ the set of all multisets over A, with the empty multiset, $\emptyset$, defined by $\emptyset(a) = 0$ for all $a \in A$.

A multiset $M : A \mapsto \mathbf{N}$, for $A = \{a_1, \ldots, a_n\}$ is represented by the state vector $w = (M(a_1), M(a_2), \ldots, M(a_n))$, w. The union of two multisets $M_1, M_2 :$ $A \mapsto \mathbf{N}$ is the addition of vectors w_1 and w_2, representing the multisets M_1 and M_2, respectively. If $M_1(a) \leq M_2(a)$ for all $a \in A$, then we say that multiset M_1 is included in multiset M_2 and write $M_1 \subseteq M_2$. A rewriting rule r over A can be defined as two multisets, (s, u), with $s, u \in A^{\#}$. A set of rewriting rules is expressed as R. A rule $r = (s, u)$ can also be represented as $r = s \to u$. Given a multiset $w \subseteq s$, the application of a rule $r = s \to u$ to the multiset w produces a multiset w' such that $w' = w - s + u$. Note that s and u can also be zero vector (i.e., empty).

3   Simulation of Natural Systems

We can model multiscale self-organizing systems such as chemical reactions, membrane systems (protocell models)[13], signal transduction systems in cells, and ecosystems using ARMS.

Belousov-Zhabotinskii Reaction

The Belouzov-Zhabotinskii (BZ) reaction displays a remarkable repertoire of complex behavior, including periodic and chaotic temporal oscillations, multiple stable stationary states, temporally and spatially periodic expanding target patterns, and rotating multi-armed spiral waves [3].

A simple abstract chemical scheme of the BZ reaction has been proposed by Prigogine and colleagues [11] in the form of the following rules:

$$A \xrightarrow{k_1} X : r_1$$
$$B + X \xrightarrow{k_2} Y + D : r_2$$
$$2X + Y \xrightarrow{k_3} 3X : r_3$$
$$X \xrightarrow{k_4} E : r_4$$

In a simulation of the Brusselator model, the frequency of applying rewriting rules follows the law of mass action, and the probabilities can be described as follows:

$$\text{Prob}(x \to x + r_1) = k_1$$
$$\text{Prob}(x \to x + r_2) = k_2\,x$$
$$\text{Prob}(x \to x + r_3) = k_3\,x^2 y$$
$$\text{Prob}(x \to x + r_4) = k_4\,x$$
$$\text{Prob}(x \to x) = 1 - (k_1 + k_2 x + k_3 x^2 y + k_4 x). \tag{1}$$

The results of these simulations are shown in Figure 1. The model exhibits oscillation between the value of X and Y (the limit cycle), where ARMS agrees well with the kinetics of the differential equation model. In this model,

$$B + X \xrightarrow{k_2} Y + D : r_2$$

plays a key role: in order to activate or inhibit this reaction, the intermediate substance Y is required, and so the rates of producing and consuming Y control the system's behaviors.

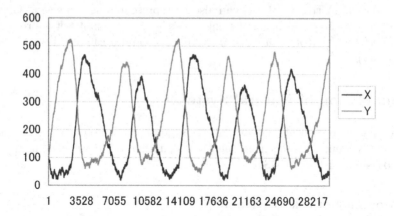

Fig. 1. Limit cycle behavior of ARMS for the Brusselator system. Parameters: $k_1 = 100$, $k_2 = 3$, $k_3 = 10^{-3}$, $k_4 = 1$.

Chemical Autopoesis

Chemical autopoiesis has been proposed by L. Luisi [8] and F. Varela [9] as representing a model for abiogenesis, and aspects of it have already been chemically realized [8], [23]. Chemical autopoiesis is a chemical system based on the surfaces of water drops in oil, where oil-soluble chemical "a" and water-soluble chemical "b" produce surfactant molecule "c" between water drops and oil; when the concentration of "c" becomes large, a water drop covered by "c" is divided into water drops of random size.

$$a + b \rightarrow c$$

In order to describe the membrane structure in ARMS we define the language MS over the alphabet $\{[,]\}$ whose strings are recurrently defined as follows:

(1) $[,] \in MS$
(2) if $\mu_1, ..., \mu_n \in MC, n \geq 1$ then $\{\mu_1, ..., \mu_n\} \in MS$
(3) there is nothing else in MS

The outermost membrane M_0 corresponds to a container, such as a test tube or reactor, that never dissolves. We describe the ARMS with the membrane as an Active Cell System (ACS); a transition of ACS is denoted by the construct:

$$\Gamma = (A, \mu, M_1, ..., M_n, R, MC, \delta, \sigma),$$

where:

(1) A is a set of objects;
(2) μ is a membrane structure (it can be changed throughout a computation);
(3) $M_1, ..., M_n$ are multisets associated with the regions 1,2,...,n of μ;
(4) R is a finite set of multiset evolution rules over A.

(5) MC is a set of membrane compounds;

(6) δ is the threshold value of dissolving a membrane;

(7) σ is the threshold value of dividing a membrane;

μ is a membrane structure of degree n, n $\geq$ 1, with the membranes labeled in a one-to-one manner, for instance, with the numbers from 1 to n. In this way, also the regions of μ are identified by the numbers from 1 to n.

(1) All the rules are applied in parallel. In every step, all the rules are applied to all applicable objects in every membrane. If there are more than two rules that can apply to an object, then one rule is selected randomly.

(2) If a membrane dissolves, then all the objects in its region are left free in the region immediately above it.

(3) All objects and membranes not specified in a rule and that do not evolve are passed unchanged to the next step.

Dissolving and Dividing a Membrane

Dissolving a membrane is defined as follow:

$$[_h a, ... [_i b, ...]_i]h \rightarrow [_h a, b, ...]_h,$$

where the ellipsis $\{...\}$ illustrate chemical compounds inside the membrane. Dissolving takes place when

$$\frac{|w_i|_{MC}}{|M_i|} < \sigma$$

where σ is the threshold value for dissolving the membrane. All chemical compounds in its region are then freed and merge into the region immediately above it.

Dissociation of Membrane Compounds

Each compound in a membrane will break off over time, and these dissipated membrane compounds are subsequently merged into the region immediately above it: for example,

$$[_0 a, b, c, c, c[_1 \mathbf{c}, c, c, a, b]_1]_0 \rightarrow [_0 a, b, c, c, c, \mathbf{c}[_1 c, c, a, b]_1]_0,$$

where membrane compound c in membrane 1 is dissipated and dissolved into the upper region. Hence in order to maintain a membrane, membrane compounds need to be continuously produced to exceed σ, otherwise the membrane will dissolve.

When the volume of membrane compounds reaches a designated threshold, then the membrane is divided. Membrane division is realized by dividing the multisets into random sizes. The frequency of membrane division is proportionate to its size as the size of a multiset becomes larger, the cell divides more frequently:

$$[_h a, b, ...]_h \rightarrow [_h a, ... [_i b, ...]_i]]_h$$

and membrane division takes place when:

$$\frac{\|w_h\|_{MC}}{\|M_h\|} > \sigma$$

where σ is the threshold for dividing the membrane. All chemical compounds in its region are then freed and separated randomly by new membranes.

Evolution of Cells

When a cell grows and reaches the threshold value for dividing, it divides into parts of random sizes. This can be seen as a mutation. If a divided cell does not have any membrane compounds, it must disappear immediately. As such, maintaining the membrane through chemical reactions inside the cell can be seen as natural selection. If a cell cannot maintain its membrane, it must disappear. Thus, both dividing and dissolving membranes produce evolutionary dynamics. Hence, all surviving cells in the ACS fit the conditions for Gnti's chemoton model [4].

Behavior of ACS

We set the ACS as $\Gamma = (A = \{a, b, c\}, \mu = \{[,]_0, ... [,]_1 00\}, M_0 = \{[a^{10}, b^{10}, c^{10}]^{100}\}, R, MC = \{c\}, \delta = 0.4, \sigma = 0.2)$ where;

(1) R, the length of the left or right-hand-side of a rule is between one and three. Both sides of the rules are obtained by sampling with replacement of the three symbols a,b, and c;

(2) Membrane structures are assumed to be $(\mu = \{[_1]1, ... [_{100}]100\})$.

step state
0. $[a^{10}, b^{10}, c^{10}]$
1. $[a^2, b^5, c^{10}]$
2. $[a^{10}, b^2, c^7]$
3. $[a^{11}, b^3, c^7]$
3. $[a^6, b^4, c^5]$

....

10. $[a^7, b^4, c^2]$

....

16. $[a^1, b^4]$

Fig. 2. An example of state transition of ACS (λ_e closes to 0.0)

λ_e Parameter

In order to investigate the correlation between the characteristics of the rewriting rule and the behavior of the model, we will introduce the λ_e parameter [15] of ARMS;

$$\lambda_e = \frac{\Sigma r_{\Delta S > 0}}{1 + (\Sigma r_{\Delta S < 0} - 1)},$$

where $\Sigma r_{\Delta S>0}$ corresponds to the number of *heating rules*, and $\Sigma r_{\Delta S<0}$ to the *cooling rules*. Heating rules take effect when the right hand side of a given rule is larger than its left hand side, while cooling rules take effect when the left hand side of a rule is larger than its right hand side. Hence, λ_e indicates the degree of reproduction of chemicals inside a cell: when λ_e approaches 0.0, the reproduction of chemicals is small, and when it approaches 1.0, reproduction is large.

step state
0. $[a^{10}, b^{10}, c^{10}]$
1. $[a^{10}, b^{10}, c^{10}]$

. . . .

91. $[a^{10}, b^{10}, c^{10}]$
92. $[[b^2], [b^3, c^1]]$
93. $[[b^2], [b^1, c^1]]$
94. $[a^{10}, b^{10}, c^{10}]$

. . . .

114. $[a^2, b^5]$
115. $[[a^1, b^2][a^1 b^3]]$
116. $[[a^1, b^2][b^2][a^1 b^3]]$

. . . .

140. $[[[a^1, b^2][b^2, c^1]][b^2][a^1 b^3][b^1, c^1]]$
141. $[[a^3, c^3][a^3, c^2][a^1 c^2]]$
142. $[b^6, c^5]$

Fig. 3. State transition of ACS (λ_e in between 0.5 and 1.0)

The behavior of the ACS is classified by this parameter. When λ_e is small, cells do not evolve but instead disappear. As λ_e becomes larger, the ACS demonstrates "cell cycle"-like behavior, wherein a small cell grows larger before dividing into smaller cells, which then in turn grow larger and divide, and so on. When λ_e exceeds 1.0, the ACS grows larger and develops a complicated internal structure, but when it exceeds 2.5, since sufficiently large membrane compounds are being produced compared to the dissolving rate of membrane compounds, the cell does not divide and simply becomes a large cell with a simple structure.

This result shows that the consumption of membrane compounds results in dynamical behavior of the system, but while sufficiently large quantities of membrane compounds remain, the system is stable and does not demonstrate dynamical behavior.

Genetic ACS, GACS

In the previous simulation, we controlled the characteristics of reactions by changing the λ_e parameter. The next step was to use reaction rules to the control the ACS, specifically by introducing heredity of reaction rules: i.e., when a cell divides, the mutated reaction rules are inherited by the divided cell. In the GACS, we denote the set of reaction rules as the rule matrix in Table 1:

step state
0. $[a^{10}, b^{10}, c^{10}]$
1. $[a^{10}, b^{10}, c^{9}]$

....

41. $[a^2, b^7, c^5]$
42. $[[b^2][b^6, c^1]]$
43. $[[b^4][b^3][b^2, c^3]]$
44. $[[b^4][b^2][b^2, c^1]]$
45. $[[b^4, c^2][b^2, c^2]][b^1, c^3]]$
46. $[[[b^3, c^1][b^1, c^1]][[b^2, c^1][b^1, c^3]]]$
47. $[[[[b^2][a^1, b^1]][b^1, c^2]][a^1, b^3, c^2]$
    $[a^1, b^2, c^3]]$

....

140. $[[[a^1, b^1][b^1, c^1]][b^2][a^1, b^1][b^1, c^1]]]$

....

214. $[[[[a^{16}, b^3, c^2][a^5, b^5, c^1][a^3, b^2]]]$
     $[a^4, b^3]][[[[b^2][b^2][a^1, b^2]][a^3, b^1]]$
     $[a^3, b^1][b^4][a^3, b^1]]]$

Fig. 4. State transition of ACS (λ_e in between 1.05 and 2.33)

step state
0. $[a^{10}, b^{10}, c^{10}]$
1. $[a^{12}, b^5, c^9]$
2. $[a^{14}, b^4, c^{10}]$
3. $[a^{13}, b^6, c^{11}]$
4. $[a^{14}, b^8, c^{12}]$
5. $[a^9, b^9, c^{10}]$
6. $[a^7, b^{10}, c^{10}]$

....

87. $[a^{17}, b^{12}, c^{21}]$

....

147. $[a^{14}, b^{20}, c^{29}]$

....

242. $[a^{17}, b^{50}, c^{44}]$

....

300. $[a^3, b^{56}, c^{58}]$

Fig. 5. State transition of ACS (λ_e more than 2.5)

Table 1. Rule matrix

	a	b	c
a	x, y_{aa}	x, y_{ab}	x, y_{ac}
b	x, y_{ba}	x, y_{bb}	x, y_{bc}
c	x, y_{ca}	x, y_{cb}	x, y_{cc}

where x, y_{ij} represents the number of x compounds of i that are transformed from the number of y compounds of j; when $x = 1$, it is not denoted. For example, $2, 3_{ab}$ is shorthand for $a, a \rightarrow b, b, b$. When the number on the left hand side of a rule (e.g., x) is one, we abbreviate it, so $a \rightarrow b, b, b$ is abbreviated as 3_{ab}.

Transmission of Reaction Rules. When a cell is divided, the reaction rules that govern the cell are copied and passed down to the new daughter cell. At that time, a point mutation occurs only in the copied rules passed down to the new cell; the initial cell retains its old rules. Point mutations occur at the cell division time point, rewrites x or y, and changes the number of transforming substances.

An Experimental Result of GACS
We set a GACS as $\Gamma = (A = \{a, b, c\}), R, \mu = \{0\}, M_0 = \{a^{10}, b^{10}, c^{10}\}, \delta = 0.4, \sigma = 0.2)$, where the R is defined as follows;

	a	b	c
a	0_{aa}	0_{ab}	1_{ac}
b	1_{ba}	0_{bb}	0_{bc}
c	0_{ca}	1_{cb}	0_{cc}.

In the evolution of the GACS, we examined the productivity of the membrane compounds of R, which is defined as the ratio of the number of membrane compounds produced to the number of non-membrane compounds: for example, denotes $a \rightarrow b, b, c, c$ and $c \rightarrow a$. Two c are produced, while a combined three of a and b are also produced: hence, this ratio for c is $Prd_{mc} = \frac{2}{3}$.

	a	b	c
a	0_{aa}	0_{ab}	1_{ac}
b	2_{ba}	0_{bb}	0_{bc}
c	2_{ca}	0_{cb}	0_{cc},

At every step of the GACS, an acs (acs_i) is selected randomly from $S_a cs$ and m_i is rewritten using r_i; the pre-defined fitness function $v_f(acs_i)$ gives the fitness value of acs_i. If this fitness value exceeds the required value θ, a sibling of acs_i is reproduced by mutating m_i and/or r_i. Otherwise, acs_i is removed from $S_a cs$ by dissolving its membrane to allow the internal compounds to be merged into the upper membrane, and a new acs, acs_i, is added to $S_a cs$ instead. The new acs_i has randomly generated reaction rules r_i and a randomly generated multiset m_i. An outline of the algorithm of the GACS is summarized below.

Figure6 illustrates the time series of productivity, where the vertical axis illustrates the steps and each dot is an R. It shows that initially, almost all Rs evolve so that $Prd_{mc} > 1$. However, after 100 steps, the productivity of the Rs decrease. At this point, both the number and size of cells increase exponentially.

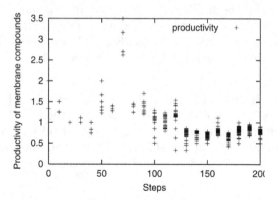

Fig. 6. Time series of productivity of membrane compounds

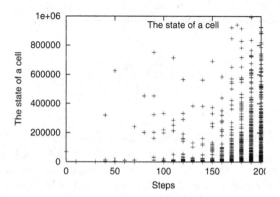

Fig. 7. Time series of the distribution of internal nodes of the whole system

Furthermore, the structure of the cells gains increasing complexity. Figure 6 illustrates the correlation between the number of cells, the size of the cells and the number of steps, where each dot corresponds to an individual cell. Figure 7 illustrates the internal nodes of the whole system. If we regard M_0 as the root and other cells as internal nodes and leaves, we can regard the whole system as a tree. We use the number of internal nodes in the tree as a metric of its complexity. Figure 7 illustrates that the number of internal nodes increases exponentially after 150 steps. It is interesting that when a cell grows into a hierarchical cell, the R evolves to have low productivity. The reason for this behavior could be that the R whose productivity is high always suffers from mutations, because it promotes membrane division and thus generates more mutations than low productivity Rs. If every cell is an elementary cell, the R must keep producing membrane compounds at a high rate. However, when the cell forms internal structure, high productivity is not necessary, because in a structured cell, if an inside cell dissolves, the cell that includes the dissolved cell obtains its membrane compounds. Therefore, as a cell evolves into a structured cell, the cell needs a high productivity of R. However, once it becomes a structured cell, high productivity of R is filtered out.

Modeling P53 Signaling Pathways

The p53 signaling network plays a major role in cell survival, as it safeguards against genetic instability, which can lead to tumor formation. However, the complicated structure of the network hampers modeling with ordinary rate equation models.

The p53 signaling network has been studied intensively because over 50 to 55% of all human cancers are reported to involve a mutation in the p53 gene [20]. The p53 protein is a transcription factor that plays a major role in regulating the response of mammalian cells to stresses and damage, mainly through the transcriptional activation of genes involved in cell cycle control (G1 arrest), DNA repair, and apoptosis [20]. In healthy cells, p53 is a short-lived and non-abundant protein, due to its rapid degradation [20]. However, in the presence of DNA damage, p53 transforms itself from a latent to an active conformation [20]. p53 has two levels of activation, depending on the level of DNA damage. Weakly activated p53 prevents damaged cells from proceeding in the cell division cycle and promotes DNA repair. Highly activated p53 induces apoptosis and eliminates mutated or irrevocably DNA-damaged cells.

To delay p53-induced apoptosis and permit cells that are not irreversibly damaged or mutated to survive, p53 forms an auto-regulatory negative feedback loop with MDM2 oncoproteins. In addition, the survival factor promotes the activation of MDM2 through PI3K-PDK1-Akt signaling and translocation of p53. Moreover, the growth factor inhibits MDM2 activation through PTEN protein and caspase activation, whereas P53 induces MDM2, which provides DNA-damaged cells the opportunity for DNA repair. Subsequently, p53 induces PTEN, which then induces the death of mutated or irrevocably DNA-damaged cells. P53 and MDM2 form a highly complex network which allows for the proliferation of healthy cells and the elimination of mutated cells [20].

In order to model these processes, we use an ACS with two membranes, one representing the nucleus and one enclosing the cytoplasm, which are labeled n and c, respectively. As such, the membrane structure is $[_c[_n]_n]_c$ where $[_c]_c$ represents cytoplasm and $[_n]_n$ represents for nucleus. The rewriting rules of p53 signaling network are given in Table2. Some rules require target membrane to be moved: for example, the rules for the "cytoplasm $([_c]_c)$", $P53 \rightarrow (P53, Nucleus)$ means that a P53 in the cytoplasm will move to "nucleus $([_n]_n)$" and instead of empty multiset, we have written "vanish" for clarity.

Biologically, MDM2 is phoshorylated by survival signaling through the PI3-kinase-PDK1-Akt pathway, which promotes rapid p53 degradation. We have summarized these interactions in the rule from R_n. Thus P53 and MDM2 from an auto-regulatory negative feedback loop. Once DND damage increases P53 is activated and trans-located from the cytoplasm to the nucleus. The activated P53 complex induces PTEM protein and caspase activation that inhibit MDM2 activation. We have summarized these interactions in the rule

$$P53 - tetrameter, DNA - damage \rightarrow P53 - tetrameter^+, DNA - damage$$

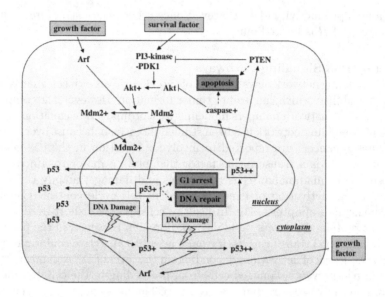

Fig. 8. Schematic model of P53 signal transduction system

Table 2. Rewriting rules of the P53 signaling network

Rules in *Cytoplasm*, R_c
    $P53 \rightarrow vanish$,
    $P53 \rightarrow (P53, \text{Nucleus})$,

Ruels in *Nucleus*, R_n
    $P53, P53, P53, P53 \rightarrow \text{P53-tetrameter}$,
    $P53 - tetrameter \rightarrow P53 - tetrameter, MDM2$,
    $P53, MDM2 \rightarrow vanish$,
    $MDM2 \rightarrow (MDM2, out)$,
    $P53 - tetrameter, MDM2 \rightarrow (P53, P53, P53, P53, Cytoplasm)$,
    P53-tetrameter DND-damage $\rightarrow P53 - tetrameter^{+}$, DNA-damage,
    $P53 - tetrameter^{+}$, DNA-damage $\rightarrow$ P53-tetrameter
    P53, MDM2 $\rightarrow$ vanish

---

from R_n where "$P53 - tetrameter^{+}$" indicates that P53-tetrameter has been activated. It forms the positive feedback loop to accelerate p53 activation. The results of the simulation correlate with biological data; when the DNA is damaged (the "abnormal state in Figure 9), a p53-tetrameter is activated, which repairs DNA damage, before returning to the normal state, where it is degraded by MDM2 into single p53. Thus, p53 and MDM2 form an auto-regulatory negative feedback loop.

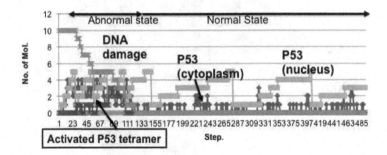

Fig. 9. Time series of concentration of P53 and MDM2 in P53 signaling network

4 Modeling Ecological System

We have modeled chemical reactions, membrane systems, and cell signaling networks. Lastly, we model a macro system - an ecosystem. An ecosystem has been reported in which plants may respond to herbivore feeding activities by producing volatile chemicals that attract carnivorous enemies of the herbivores [18]. These volatiles are not merely the result of mechanical damage, but are produced by the plant as a specific response to herbivore damage. We model this tritrophic system by using an ARMS with stochastic transition. Let the symbol "a" be a leaf, "b" be a herbivore, "d" be a carnivore and "c" be the density of herbivore-induced volatiles. Furthermore, we define "e" to be an empty state in order to introduce the death state. A plant is defined implicitly as a number of leaves. Evolution rule R_1 is defined as follows:

$$a \xrightarrow{k_1} a, a \quad r_1 \quad (increase\ in\ the\ number\ of\ leaves),$$

$$a, b \xrightarrow{k_2} b, b, c \quad r_2 \quad (herbivore\ eats\ a\ leaf),$$

$$d, b, c \xrightarrow{k_3} d, d \quad r_3 \quad (carnivore\ catches\ a\ herbivore),$$

$$d \xrightarrow{k_4} e \quad r_4 \quad (death\ of\ a\ carnivore),$$

$$b \xrightarrow{k_5} e \quad r_5 \quad (death\ of\ a\ herbivore).$$

Rule r_1 corresponds to the sprouting and growth of a plant, r_2 corresponds to a herbivore eating a leaf and the leaf generating volatiles, r_3 to the herbivore being preyed upon by a carnivore, r_4 to the death of a carnivore, and r_5 to the death of a herbivore. More precisely, r_2 denotes the case when a leaf (a) exists, and a herbivore (b) eats the leaf and reproduces. The leaf produces volatile compounds (c) upon being eaten, to attract carnivores. Rule r_3 denotes the case when there is a herbivore (b) present with volatiles (c), and a carnivore (d), attracted by the volatiles of r_2, catches the herbivore and reproduces (d, d).

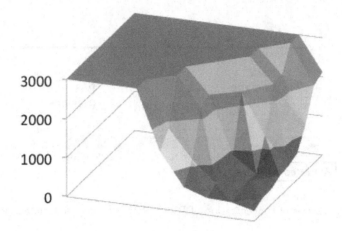

Fig. 10. Symbiotic relation of plant, herbivore and carnivores, horizontal axes illustrate k_1 and k_2 and the vertical axis illustrates step time of sustaining symbiotic relations; (with herbovire-induced plant volatiles HIPV)

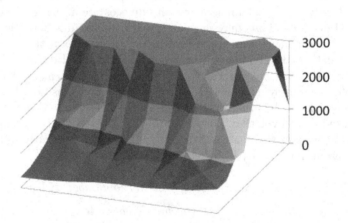

Fig. 11. Symbiotic relation of plant, herbivore and carnivores, horizontal axes illustrate k_6 and k_7 and the vertical axis illustrates step time of sustaining symbiotic relations; (without HIPV)

Using this model, we compared the outcomes of systems where leaves generate volatiles against those of systems where leaves do not. The evolution rules of the system without volatiles R_2 are defined as follows:

$$a \xrightarrow{k_6} a, a \quad k_6 \quad (increase\ in\ the\ number\ of\ leaves),$$

$$a, b \xrightarrow{k_7} b, b \quad k_7 \quad (herbivore\ eats\ a\ leaf),$$

$$d, b \xrightarrow{k_8} d, d \quad k_8 \quad (carnivore\ catches\ a\ herbivore),$$

$$d \xrightarrow{k_9} e \quad k_9 \quad (death\ of\ a\ carnivore),$$

$$b \xrightarrow{k_{10}} e \quad k_{10} \quad (death\ of\ a\ herbivore).$$

In both R_1 and R_2, the symbiotic relationship between plants, herbivores and carnivores can be identified. It is interesting that R_2 (the relationship under a system without volatiles, Figure 11) is more likely to collapse than R_1, under the same conditions, with the rate of collapse dependent on the number of plants, herbivores and carnivores in the initial state, and the reaction rate of r_3 and r_8.

On the other hand, it shows that a system with volatile generation by plants (Figure 10) is more robust than a system without it under otherwise identical conditions. Thus, the role of volatiles warrants further investigation.

5   Dissipation of Information

By modeling self-organizing systems at such different scales-a BZ reaction, the p53 signaling system, and a chemical ecosystem-with ARMS, we have discovered that these models have a common algorithmic structure. In every model, interactions produce an "intermediate substance" that is consumed by evoking another interaction. In the BZ reaction, the intermediate product "Y" is produced by the reaction of "A, X Y" and "Y" is consumed by evoking the auto-catalytic reaction "X, X, Y $\rightarrow$ X, X, X". In the protocell system, membrane compounds "C" are produced by the reaction of "A, B C" and dissipation of "C" , requiring the generation of another "C" in order to maintain the membrane (hence, dissipation evokes mobilization of "C"). In the p53 signaling system, the intermediate product "$p53 - tetrameter^+$" is produced by damaged DNA and consumed by repairing damaged DNA, a reaction that produces another intermediate product "p53-tetrameter", which in turn is consumed by generating a third intermediate product, "MDM2", through the reaction of "p53-tetrameter $\rightarrow$ p53-tetrameter, MDM2". In the ecosystem, volatile chemical "c" is produced by the feeding activity of herbivores, as described as the reaction of "a,b $\rightarrow$ b,b,c", and is consumed by evoking the carnivore's feeding on herbivores "d,b,c $\rightarrow$ d,d."

Table 3. Dissipation of information in natural systems

BZ reaction X,X,**Y** $\rightarrow$ X,X,X
Proto cell   membrane compound $\rightarrow$ **dissolved**
P53 system $P53 - tetrameter^+$, **DNA damage** $\rightarrow$ $P53 - tetrameter^+$
Ecosystem   carnivore, herbivore, **chemical** $\rightarrow$ carnivore, carnivore

As summarized in Table 3, in natural systems, producing and consuming intermediate substances are important behaviors for controlling self-organization. These intermediate substances have not been prepared and are generated through interactions and consumed by evoking other reactions. As such, these intermediate substances can be regarded as "information", and they regulate self-organization phenomenon. We call these information producing and consuming interactions "dissipation of information." This is widely realized in natural systems, and so there are opportunities to manipulate natural systems through this mechanism.

6 Harness the Nature for Computation

Using this concept of dissipation of information, intermediate substances do not fully control the system, but only partly, and the internal dynamics of individual substances influence the dynamics of the whole system via a ripple effect. For example, in order to control a group of sheep, shepherds do not grip each individual sheep but threaten several sheep by voice or by using sheepdogs. This intimidation can be regarded as an "intermediate substance": it is consumed through affecting the internal dynamics of a group of sheep, likely gathered together, as when one escaping sheep leads other sheep to follow. We refer to this indirect control of internal dynamics through affecting part of a system as "harness", and those points where internal dynamics can be affected as "harness points."

Harness a System by Dissipative of Information
We harness a food chain model which is composed of nine species of Lotoka-Volterra equations. Before harnessing, the model is denoted as;

$$a_0, a_1 \quad \rightarrow \quad a_1, a_1,$$
$$a_1, a_2 \quad \rightarrow \quad a_2, a_2,$$
$$\cdots\cdots\cdots$$
$$a_9, a_0 \quad \rightarrow \quad a_0, a_0,$$

We set initial species populations as 20, 30, 40, 50, 60, 70, 80, 90, 100, then ran the system through 190 time steps. Species whose initial populations were 90, 70, 50, and 30 become extinct (Figure6). We harness the system by using

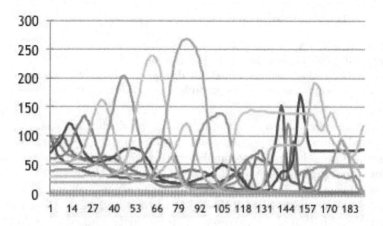

Fig. 12. Population dynamics of the food chain which is composed of 10 species of Lotoka-Volterra model

information dissipative through introducing intermediate substances $c_0, ..., c_9$ as harness points;

$$a_0, a_1, c_0 \quad \rightarrow \quad a_1, a_1, c_1,$$
$$a_1, a_2, c_1 \quad \rightarrow \quad a_2, a_2, c_2,$$
$$............$$
$$a_9, a_0, c_9 \quad \rightarrow \quad a_0, a_0, c_0,$$

where at each food chain step, an intermediate substance is consumed and produced, hence oscillations become very small (Figure 6).

After introducing harness points, although the distribution of initial populations remained the same, large amplitude oscillations were inhibited and no species become extinct (Figure6).

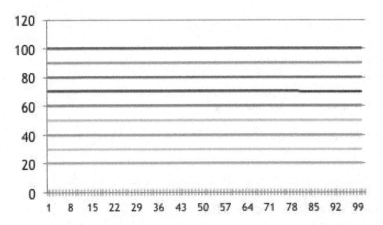

Fig. 13. Harnessed population dynamics of the food chain which is composed of 10 species of Lotoka-Volterra model with intermediate substances. It is noticed that, scale of the vertical axis is different from Figure6.

GPA
We harness GACS to solve a problem and named it the Genetic Protocell Algorithm (GPA). We solve the *doubling problem*, calculating the sum of a and b then showing the result as a number of c ($c = 2(a + b)$); The construct of GPA Γ^{GPA} is

$$\Gamma^{GPA} = (S_{acs}, f_s, M_s, \theta),$$

where

(1) $S_{acs} = \{acs_1, ..., acs_n\}$ is a set of ACS;
(2) $f_s(acs_i)$ is a function which gives the fitness value of acs_i;
(3) $M_s(acs_i)$ is a operation which mutates acs_i;
(4) θ is the threshold of fitness value for natural selection;

In the initial state, we set 100 elementary cells inside M_0. No compounds are transformed among protocells and no input and output are assumed.

The way of dissolving and dividing are the same as in GACS. After n rewriting steps, if the number of c in an acs is smaller than 7, its membrane is dissolved (a harness point), while if the number of c is larger than 9, the membrane is divided and a point mutation R is passed down to the new membrane. When a membrane is divided, chemicals inside the divided cell and its parent cell are reset to $\{a^2, b^2, c^0\}$ and the same problem again is solved again. The algorithm of GPA is summarized as follows;

begin
while (flag = NULL)
select $acs_i \in S_{acs}$ randomly
rewriting $m_i \in asc_i$
obtain the fitness value $v_f = f_s(m_i)$
if $(v_f < \theta)$
then
$S_{acs} - acs_i$
generate a new acs_i randomly
$S_{acs} \cup acs_i$
else
a generate new acs, acs_j by mutating
$M \cup acs_j$
if $\exists acs_k \in S_{acs}$ convergence
then flag:=T
end.

An Experimental Result of a GPA

The GPA in this experiment was defined as follows; $\Gamma^{GPA} = (S_{acs}, f_s = M(c), c \in acs_i(1 \geq i \geq 100), \theta = f_s(acs_i) < 7)$, where $acs_i(1 \geq i \geq 100)$ is defined as follows;

$$\Gamma = (A = \{a, b, c\}, \mu, \mu = \{[,]_0, ...[,]_{100}\}, M_0 = \{[a^2, b^2, c^0]^{100}\}, R, MC = \{\})$$

and θ is the same as previous section and σ is not used. In the initial state, R is After 5,000 reaction steps, every R which reached the solution within 8 steps was

	a	b	c
a	0_{aa}	0_{ab}	1_{ac}
b	1_{ba}	0_{bb}	0_{bc}
c	0_{ca}	1_{cb}	0_{cc},

selected; Chemical components of selected cells were reset to the initial state, $[a^2, b^2, c^0]$ and the calculations were performed again. Next, Rs that solved the problem within 5 steps were selected. In this case, they had converged into the same reaction rules as follows; this R is closed to the result; By harnessing GACS, we attempted to treat a system as a living thing, whereby in order to obtain the desired result, we observed its output (behaviors) and harnessed them by changing the environment without halting computation; by harnessing, natural selections and mutations lead to the GACS that achieves our goal.

$$
\begin{array}{llll}
a \to & a,c, & a \to & a,b,c,c, \\
b \to & a,b,c,c, & b \to & a,b,c,c, \\
c \to & c,c; & c \to & c,c;
\end{array}
$$

$$
\begin{array}{ll}
a & \to a,c,c, \\
b & \to b \\
c & \to c,c;
\end{array}
$$

$$
\begin{array}{l}
a \to a,c,c, \\
b \to c,c, \\
c \to c,c;
\end{array}
$$

While the concept of a harness has the concept been used, but not named as a "harness," its realizations have already been used in various fields, as described in the following examples.

Spray Treatment. Secretory otitis media (SOM), persistent fluid in the middle ear cavity, stems from an unknown cause. S Skovbjerg et al. [12] used a nasal spray containing alpha-streptococcal bacteria considered to have a protective effect for children prone to middle ear infections, or otitis media. They investigated the clinical, bacteriological, and immunological effects of treatment with pro-biotic bacteria on SOM.

In the double-blind pilot/preliminary study, 60 children with chronic SOM (median 6 months) scheduled for insertion of tympanostomy tubes were assigned randomized nasal spray treatments with Streptococcus sanguinis, Lactobacillus rhamnosus, or placebo for 10 days before surgery. Clinical evaluation was carried out after 10 days of treatment. Middle ear fluid (MEF) was collected during surgery for quantification of cytokines and detection of bacteria by culture and polymerase chain reaction. Nasopharyngeal swabs were obtained before treatment and at surgery.

Complete or significant clinical recovery occurred in 7/19 patients treated with S. sanguinis compared to 1/17 patients in the placebo group ($p < 0.05$). In the L. rhamnosus treatment group, 3/18 patients were cured or became much better ($p = 0.60$; compared to placebo). Spray treatment did not alter the composition of the nasopharyngeal flora or the cytokine pattern observed in the nasopharynx or MEF, except for a higher level of IL-8 found in the nasopharynx of L. rhamnosus-treated children. This study shows that spray treatment with S. sanguinis may be effective against SOM.

Harness Point. In the spray treatment, the internal dynamics is the population dynamics between the causative microorganism and the resident microbiota and a harness point is to support the resident microbiota by the nasal spray.

$$a \to c,c,$$
$$b \to c,c,$$
$$c \to c.$$

Pest Control Technique. In the previous section, we modeled an ecosystem where plants responded to herbivore feeding activity by producing volatiles that attracted carnivorous enemies. J. Takabayashi et al. have used this model as harness for a pest control technique [22]. They found that cruciferous plants infested by diamondback moth larvae emitted a blend of volatiles that attract Cotesia vestalis, carnivorous braconid wasps that are natural enemies of the larvae. They also conducted experiments using a synthetic blend of C. vestalis attractants for diamondback moth control. They have identified the braconid inducement material triggered by the diamondback moth, and have developed an action control technique for the attracted wasp and another technique for activating the induced wasp.

Harness Point. In the pest control, the internal dynamics is the information dissipative in the population dynamics of ecosystem and a harness point is intermediate substance, chemical volatiles, HIPV.

7 Conclusion

We have considered the harness from the viewpoint of natural computing. We investigated facsimile computational models of spontaneous self-organization in nature, and identified dissipation of information flow as a common mechanism. Since the dissipation of information algorithm only partly affects a system's internal dynamics, we consider it an indirect method of control, and conceived the concept of harnessing natural systems for the harness. We discovered that this concept is already used in harnessing as summarized in Table 4.

Table 4. Examples of different kinds of harness based on the concept of a harness, classified according to synthesis method, harness point, and materials

class	Primary	Harness point	Secondary
Wetware	Microbiology	population dynamics	Spray treatment
	Chemical Ecology	Dissipation of Info.	Pest Control
Software	Simulation of	Dissipation of Info.	Genetic Proto-cell
	Proto-cell	and population dynamics	Algorithm

Roughly, we can have two types of assumptions of the world: the digital world assumption (DWA) and the analog world assumption (AWA). In the DWA, every substance has a digital existence (i.e., either exists or does not exist) while in the AWA, every substance has an analog existence. Hence, in the DWA, we can use proof by contradiction, a powerful and convenient tool for deduction; however, we cannot use it in the AWA. Computing, although usable in the DWA, is an even more important tool within the AWA.

Nature is composed of gray boxes: in many cases, we can partly know about what is inside a box but not completely know about it. We describe a world composed of gray boxes as the oracle world, and a world composed of white boxes as the constructive world.

In computer science, we have mainly designed algorithms in the constructive world with the DWA; however, we can not compose algorithm by using gray boxes but white boxes. Hence we propose a constructive method as harness; science for the harness requires control methods in the oracle world with the AWA. Natural computing will be able to offer this through the paradigm of harnessing nature for computing. ·

In this position paper, we propose the concept of a harness and show that it has already been utilized for creating the harness. In order to better understand the harness concept, basic principles must be identified by idealizing it. Harnessing has nonetheless been used heuristically, and here we have shown examples of harnessing through dissipative information, a technique that could be used for multiscale phenomena such as drug design, environment conservation, and consensus making in a society.

Acknowledgments. This work is supported by JSPS KAKENHI Grant Numbers 23300317 and 24520106 and the Grant-in-Aid for Scientific Research on Innovative Areas Grand Number 2404002.

References

1. Dittrich, P., Ziegler, J., Banzhaf, W.: Artificial chemistries, a review. Artif. Life 7(3), 225–275 (2001)
2. Jain, S., Krishna, S.: A model for the emergence of cooperation, interdependence, and structure in evolving networks. Proc. Nat. Acad. Sci. 98(2), 543–547 (2001)
3. Field, R.J., Burger, M.: Oscillation and Traveling Waves in Chemical Systems. Willey, New York (1985)
4. Ganti, T.: Organization of chemical reactions into dividing and metabolizing units: the chemotons. BioSystems 7, 15–21 (1975)
5. Gillespie, D.T.: A General Method for Numerically Simulating the Stochastic Time Evolution of Coupled Chemical Reactions. J. Comp. Phys. 22, 403–434 (1976)
6. Kakimoto, K., Taura, T.: A framework for analyzing sustainability by using the rewriting system. In: Pro. Intern. Symp. on Environmentally Conscious Design and Inverse Manufacturing, vol. 3, pp. 69–74. IEEE (2003)
7. Kakimoto, K., Shiose, T., Taura, T.: Analyzing Sustainability of Circulatory System by Using the Rewriting System. In: Proceedings of EcoDesign: 2nd International Symposium on Environmental Conscious Design and Inverse Manufacturing, pp. S103–S107. IEEE (2001)

8. Luisi, P.L.: Defining the transition to life: self-replicating bounded structures and chemical autopoiesis. In: Stein, W., Varela, F. (eds.) Thinking About Biology. Addison-Wesley, New York (1993)

9. Luisi, P.L., Varela, F.: Self-replicating micelles: a chemical version of minimal autopoietic systems. Origins Life Evol. Biosphere 19, 633–643 (1989)

10. International Journal of Natural Computing. Springer

11. Nicolis, G., Prigogine, I.: Exploring Complexity, An Introduction. Freeman and Company, San Francisco (1989)

12. Skovbjerg, S., Roos, K., Holm, S.E., et al.: Spray bacteriotherapy decreases middle ear fluid in children with secretory otitis media. Arch Dis. Child (August 19, 2008), doi:10.1136/adc.2008.137414

13. Suzuki, H.: An Approach to Biological Computation: Unicellular Core-Memory Creatures Evolved Using Genetic Algorithms. Artificial Life 5(4), 367–386 (2000)

14. Suzuki, Y., Tsumoto, S., Tanaka, H.: Analysis of Cycles in Symbolic Chemical System based on Abstract Rewriting System on Multisets. In: Proceedings of International Conference on Artificial Life V, pp. 482–489. MIT Press (1996)

15. Suzuki, Y., Tanaka, H.: Order parameter for symbolic chemical system. In: Adami, C., et al. (eds.) Artificial Life IV, pp. 130–142. MIT Press (1998)

16. Suzuki, Y., Tanaka, H.: Chemical evolution among artificial proto-cells. In: Artificial Life VII, pp. 54–64. MIT Press (2000)

17. Suzuki, Y., Fujiwara, Y., Takabayashi, J., Tanaka, H.: Artificial Life Applications of a Class of P Systems: Abstract Rewriting Systems on Multisets. In: Calude, C.S., Pun, G., Rozenberg, G., Salomaa, A. (eds.) Multiset Processing. LNCS, vol. 2235, pp. 299–346. Springer, Heidelberg (2001)

18. Suzuki, Y., Takabayashi, J., Tanaka, H.: Investigation of tritrophic interactions in an ecosystem using abstract chemistry. J. Artif. and Robot. 6(3), 219–223 (2001)

19. Suzuki, Y., Davis, P., Tanaka, H.: Emergence of auto-catalytic structure in stochastic self-reinforcing reaction networks. J. Artif. and Robot. 7, 210–213 (2003)

20. Suzuki, Y., Tanaka, H.: Modeling P53 signaling network by using multiset processing. Applications of Membrane Computing Series: Natural Computing Series, pp. 203–215. Springer, Tokyo (2006)

21. Suzuki, Y.: An investigation of the Brusselator on the mesoscopic scale Inter. J. of Parallel, Emergent and Distributed Sys. 22(2), 91–102 (2007)

22. Urano, S., Uefune, M., Takabayashi, J.: Inspection example of the diamondback moth prevention - effect with natural enemy attractant "bee cool". In: 18th, Special Interest Group meeting of Natural Enemy Usage, The Japanese Society of Applied Entomology & Zoology (2008)

23. Walde, P., Goto, A., Monnard, P.A., Wessicken, M., Luisi, P.L.: Oparin's reactions revisited: enzymatic synthesis of poly(adenylic acid) in micelles and self-reproducing vesicles. J. Am. Chem. Soc. 116, 7541 (1994b)

Things Theory of Art Should Learn from Natural Computing

Fuminori Akiba

Graduate School of Information Science
Nagoya University
akibaf@is.nagoya-u.ac.jp

Abstract. In this paper, we first depict a short history of arts and make clear that current critical artworks are not successor to original fine arts. Then, we acknowledge that some key points of fine arts remain in natural computing, especially in the idea of 'harnessing.' Finally, we re-evaluate artworks from the point of view of natural computing, especially from the idea of 'harness' in natural computing. We further suggest the possibility of making a new lineage, for example, from horticulture in the 18th century through land art to the idea of harnessing in natural computing, and to expect new successors to fine arts in extension of this lineage.

Keywords: theory of art, natural computing, harnessing, history of art.

1 Introduction

The idea of natural computing gave me a shock because there I saw a phantom of fine arts that was already lost. They forced me to reconsider what fine arts originally were and where they are now. In my presentation below, I would like to make clear the following: 1) current critical artworks are not successors to original fine arts; 2) rather, the idea of natural computing shares key points with the original idea of fine arts. 3) In addition, it is possible to reevaluate artworks from the point of view of natural computing. Through this presentation, I would like to show that the theory of art must learn from natural computing.

2 Critical Artworks

2.1 Critical Artworks Irritate Me

Artworks irritate us because they do not satisfy our curiosity; rather, they force us to find lessons to learn, in other words, they force moral knowledge in us. Take Cannacher's *Addict to Plastic* for example. I saw a still of this video at the exhibition of Ars Electronica in 2010. The still consisted of two things: a heap of trash and a TV set frame. The frame of the TV set lost its tube, so the heap of trash could be seen through the frame.

Y. Suzuki and T. Nakagaki (Eds.): WSH 2011 and IWNC 2012, PICT 6, pp. 71–81, 2013.

As usual, it irritated me because it made me think of the meaning of this work. I imagined that the frame of the TV set is the symbol of mass media. In its lifetime, this TV set showed us glittering fictions through its tube. But now, it lost its tube and reveals the real world, that is, a heap of trash. The frame itself is a part of a heap of trash. Therefore, the lesson which we must learn from this artwork might be, for example, 'do not trust the mass media, it is trash, and it conceals the truth behind its tube.' However, in comparison to the title of this work, "Addict to Plastic," my interpretation seems to miss the point. So I am forced to watch the whole video and to reconsider what the lesson of this artwork is. That irritated me again.

Nevertheless, whatever the lesson may be, it is certain that we regard criticism as one of the important functions of artworks. Artworks reveal the invisible truth to us.

2.2   Where Are the Fine Arts?

Frankly speaking, I feel disgusted with the flood of such critical artworks. People say artworks tell the truth we do not know, but is it true? I do not think so. More than fifty years ago, Beardsley (1958) logically negated artistic truth. What are we so ignorant about that we need to receive lessons from artists? Of course, sometimes we can learn something from artworks, but we can learn more lessons from other sources without visiting museums. Finally, should artworks be critical? My answer is 'No.' I believe such critical artworks completely differ from the original concept of 'fine arts.'

Then, is the legacy of fine arts completely lost? Is there any possibility to find it? In order to answer these questions, we turn our eyes to natural computing because some of the key points of the original concept of fine arts seem to remain in natural computing. I think theory of art can learn something from the idea of natural computing, especially from the idea of 'harnessing,' which natural computing proposes.

In the next chapter (Chapter 3), we look back to a short history of arts with the help of previous studies. Through this short history, we would like to extract some of the key points of the notion of 'fine arts' and to confirm the fact that critical artworks deviate from fine arts (Chapter 4). Then, we point out that natural computing shares some of the original key points of fine arts. Finally, we would like to find another lineage of artworks that has been overlooked so far (Chapter 5).

3   A Short History of Arts

3.1   Traditional Distribution of Arts

Probably we all know about the traditional distribution between liberal arts and mechanical arts. Hence, I would like to begin my talk with this distribution (Figure 1).

liberal arts   -   mechanical arts
[intellectual]   [manual][repetitive]

Fig. 1. Traditional distribution of arts

The Encyclopedia by Diderot and d'Alembert (1751) introduces the reason why people distribute arts into these two categories: liberal arts are produced more by the mind than the hands, while mechanical arts are produced more by the hands than the mind (p.714). The latter arts are manual and must be repetitive. Though Diderot and d'Alembert criticize this idea because it results in false disdain for manual labor, we can accept this division as a starting point.

3.2 Reorganization of the Distribution in the 18th Century: Machine Arts and Fine Arts

According to Otabe (2001), in the 18th century, this traditional division was forcedly reorganized by the appearance of new kinds of arts (pp.4-14). They are 'machine [*machinal* in French]' arts and 'fine arts [*beaux-arts* in French].'

The Preliminary Discourse to the Encyclopedia by d'Alembert (1751) explains the reason why they are called 'machine [*machinal*]': they automatically operate by themselves without dependence on human hands [*operations purement machinales* in contrast to *operation manuelle*] (p.xiij). This independence from human hands Otabe (2001) calls 'dis-habituation' of arts (p.11) and sees in this dis-habituation a germinate tendency to technology. Figure 2 shows the relationship between the traditional division and machine arts.

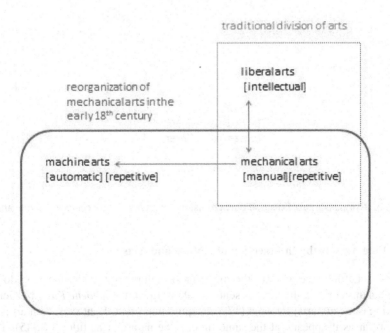

traditional division of arts

liberal arts
[intellectual]

reorganization of
mechanical arts in the
early 18th century

machine arts
[automatic] [repetitive]

mechanical arts
[manual] [repetitive]

Fig. 2. Reorganization of traditional division of arts in the 18th century (1): machine arts

In the 18th century, 'fine arts' were recognized as new kinds of arts. Included in fine arts were music, poetry, painting, sculpture, and dance. Concerning fine arts, there are at least three important points (Battuex 1747, pp.27-28, d'Alembert 1751, p.xxij): 1) fine arts have 'pleasure' [*plaisir, agrément*] for their object; 2) fine arts are the imitation of nature (Battuex 1747, d'Alembert 1751, see also Panofsky 1968); 3) according to d'Alembert (1751), fine arts belong to liberal arts and the practice of fine arts principally consists in an 'invention' of genius. Otabe (2001) sees another 'dis-habituation' in fine arts (pp.12-14): the rules of fine arts are invented by geniuses and hence cannot be produced by routine habit. Therefore, the rules of fine arts, though they are called 'rules,' seem unforeseeable for ordinary persons with finite intelligence. In this sense, fine arts are likened to having a complex nature; they also make us feel the infinite beyond us. Anyway, in all three points, fine arts were different from previous arts. This urged the reorganization of arts. (Figure 3).

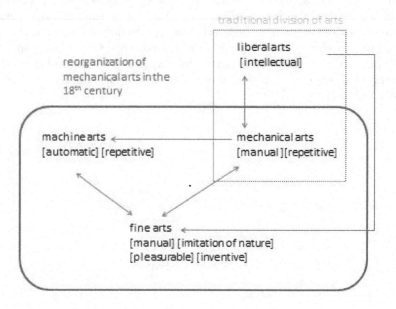

Fig. 3. Reorganization of the traditional division of arts in the 18th century (2): fine arts

3.3   Fine Arts as the Intersection of Nature and Arts

As we saw before, fine arts are the imitation of nature. But, what and how do they imitate? In this context, the famous sentence about fine arts [*Schoene Kuenste,* German translation of beaux-arts] by Kant (1790) is quite important. Kant says, "An art is fine art, in so far as it appears, at the same time, to be nature" (the title of SS45). To be nature means that it must not seem to be an artifact. Then, how can arts appear to be nature? The answer is as follows: If we feel pleasure [and purposiveness] from arts without awareness of the maker's intention and of their simple accordance with artificial rules, then they could appear to be natural and hence called fine arts. On the contrary, if we see the maker's intention in arts and feel that they are apparently subject to artificial rules that were established beforehand, such arts can never be fine arts, they are only mechanical arts.

Therefore, it seems quite natural that many thinkers (Shaftesbury, Addison, Burke, Kant, etc.) in the 18th century mentioned horticulture, the so-called English garden (see Otabe 2001, p.148ff.). 'English taste in gardens,' in contrast to the artificiality and 'stiff regularity' of French gardens, 'pushes the freedom of the imagination' because of their complexities (in Kant's words, 'nature which is extravagant in its varieties to the point of opulence,' SS22). Horticulture is at the intersection between nature and arts, and it was thought to be representative of fine arts.

3.4   Summary of This Chapter

From our discussion above, we can extract the following as the key points of fine arts (Figure 4): 1) fine arts are different kinds of arts that caused the reorganization of the traditional division of arts; 2) the principal object of fine arts is pleasure; 3) fine arts are the imitation of nature in its complexity; and 4) they are products of invention and make us feel the infinite.

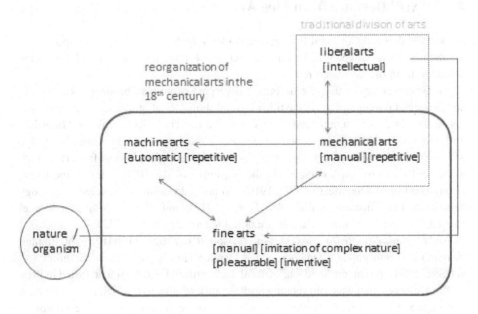

Fig. 4. Summary of Chapter 3

3.5   Important Note (1)

Though in the 18th century's discourse the rules of fine arts are said to be unforeseeable, the makers of fine arts completely grasp what these rules are. Of course, in the process of making a work of art, there are the moments when the unforeseeable invention suddenly comes to the maker. For example, in making innumerable sketches, Raphael suddenly found a completely new, but optimal composition for the altar. However, it does not mean Raphael made his work without knowing what he did (Wendler 2008). Generally speaking, since the object of fine arts is pleasure, the maker must design his/her works so that the viewers surely experience pleasure (if not, the maker loses his/her job). Even when the painting has multiple meanings (open-ended), the maker must carefully design it so that the painting would surely have multiple meanings (open- ended). Multiple meanings must never be the result of a haphazard job. Therefore, in turn, if an art, which is made with deliberate design, seems to be natural, in other words, if the viewer feels the splendidness of it but does not perceive the traces of deliberate design, the art belongs to fine arts and brings the viewer pleasure. In case of horticulture, we can say the same thing. The maker of fine arts is responsible for the result of his product.

4   "Art" Deviates from Fine Arts

As we saw above, the original fine arts are completely different from contemporary art we mentioned in Chapter 2. In this chapter, we briefly trace when and how such deviation from fine arts occurred.

Everyone can agree that the definitive diverging point was brought about by the appearance of photography. As the title of an exhibition catalogue says, photography is not 'a new art,' but 'another nature' (Kunsthalle der Hypo-Kulutstiftung Muenchen 2004). Specifically, the appearance of sequence photographs of Eadweard Muybridge and Etienne-Jules Maley turned artists away from the task of traditional fine arts, which is the imitation of complex nature. In the beginning of the 20th century at the latest, Futurists released a manifesto (1909, 1910) and praised machines, depicted technology beautifully like Giacomo Balla's *Arc Lamp* (1913), and Richard Mutt aka Marcel Duchamp failed to exhibit a ready-made urinal named *Fountain* (1917).

Andy Warhol's *Brillo Boxes and Campbell's Boxes* (1964) might imitate industrialized and popularized society and Damien Hirst's *Away from the Flock* (3rd version, 1994) might criticize our natural and artificial environment (see Godfrey 1998). However, they have no relation with the task of fine arts, which is the imitation of complexity in nature. They are 'against nature' or 'against science and technology.' (Figure 5)

from the 19ᵗʰ century to the 20ᵗʰ century

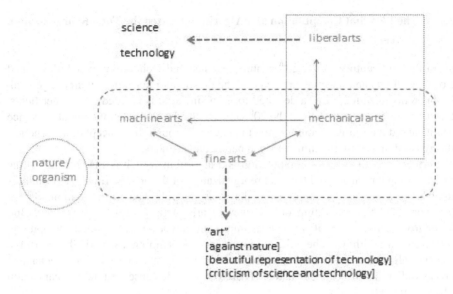

Fig. 5. Art deviates from fine arts

Photography forced us to rethink the division between machine arts and fine arts. Furthermore, it brought radical change to the concept, 'scientific objectivity' (see Daston & Galison 2007). On the contrary, art is isolated both from other fields of art and from nature. Even if we add more works such as Sherry Levine's *Fountain* (1989) to the list of art, nothing happens. It is only a comment to Duchamp's precedent within the closed area called 'artworld.'

Someone might argue that Balla, for example, tried to depict the speed and complex movements of swallows (1913), so art does not lose its relation to nature. Even if so, however, his paintings had no power to cause reorganization between arts. They only followed the formal appearances of movements, which were already recognized in photographic sequences.

5   What and How Should We Evaluate? Lessons from Natural Computing

In this chapter, we will try to point out the following: natural computing shares some key points of fine arts, though, of course, it is not fine arts, but science. Then, we would like to suggest, from the point of view of natural computing, what we should consider as possible successors to fine arts.

At the end of chapter 3, we already extracted the following as key points of fine arts (see Figure 4): 1) fine arts are different kinds of arts that caused the reorganization of the traditional division of arts; 2) the principal object of fine arts is pleasure; 3) fine arts are the imitation of nature in its complexity; and 4) they are products of invention and make us feel the infinite. We begin with the first key point.

5.1 The Idea that Computation and Algorithm Caused the Total Reorganization of Arts

From the 19th century to the 20th century, science and technology, which developed from liberal arts and machine arts, caused the drastic reorganization of arts in general. In this reorganization, 'art,' a deviated form of fine arts, is isolated from other fields. Furthermore, in the latter half of the 20th century, the development of computer science again caused the reorganization of arts in general. Through the concept of cybernetics, it surpassed in a sense the limit between nature and artifacts.

Furthermore, the idea of computation and algorithm, which is fundamental to the idea of computation, caused the total reorganization of the arts. Normally, algorithm is thought as a process or set of rules that must be followed in calculation. So we misunderstand that algorithm only relates to arithmetic or computer science. But, Yasuhiro Suzuki, one of the most important advocates of natural computing, reinterprets algorithm as 'the order of movement and its timing' (Suzuki 2012, p.5). We can find algorithm as such everywhere: in liberal arts, machine arts, and mechanical arts as well as in artifact and in nature. Therefore, it could cause the total reorganization of arts (Figure 6).

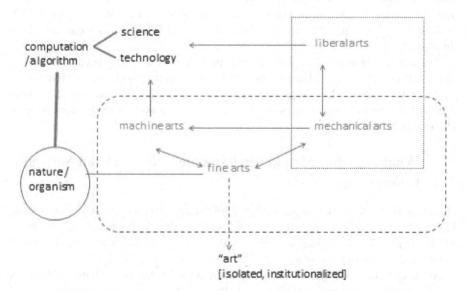

Fig. 6. The idea that computation and algorithm caused the total reorganization of arts

5.2   The Idea of 'Harness' in Natural Computing Imitates Nature in Its Complexity

It is quite clear that natural computing tries to imitate nature. According to the website of natural computing, one of the missions of natural computing is to 'explore computational processes observed in nature, and human-designed computing inspired by nature' (Springer's website).

Furthermore, the idea of 'harness' in the sphere of natural computing clearly shows that natural computing tries to imitate nature in its complexity. In the original sense, the word 'harness' means "a set of strips of leather and metal piece that is put around the horse's head and body so that the horse can be controlled and fastened to a carriage, etc." (*Oxford Advanced Dictionary*). In natural computing, harness denotes an alternative means of control. Suzuki (2012) explains it by way of comparison with the traditional method of control (pp.7-8). Imagine a shepherd whose task is to move a flock of sheep from here to a pen. In order to control a flock and achieve his task, the shepherd has two alternatives. One is to hold each sheep directly and individually move each sheep to the pen. The other is to walk behind a flock of sheep only with a picture of a wolf in his hands. The former is direct control. The latter is 'indirect' control, and it is this indirect type of control that natural computing calls 'harness' and tries to imitate.

But, why do we say that to imitate this indirect type of control, that is, harnessing, is to imitate the complexity in nature? We answer this question in the following manner: just as a shepherd, who does not know the entire mechanism of such indirect control, can move a flock –a complex natural system— only by using one picture –a minimum artifact—, so natural computing operates a complex natural system by using minimum artifacts. For example, Shinya et al. (2012) report that they found the possibility to utilize the data of synchronized pattern of functional gene clusters –minimum artifacts—by harnessing the disrupted condition of the influenza virus towards normal homeostatic flow –a highly complex natural system (p.2).

If we can apply the distinction between 'constructive' computation—of which we can explicitly show every step— and 'oracle' computation—of which we cannot explicitly show every step—(see Suzuki (forthcoming)) to the definition of harnessing, then we say that 'harnessing' provides an alternate means of control which operates 'oracle' computation by using the products of 'constructive' computation. It concerns complexity –which people in the 18th century called 'the infinite' – in nature.

5.3   Important Note (2)

In order to preclude misunderstandings, here we emphasize that scientists must have good prospects of the result of harnessing if they do it. If someone projects any artifact into a natural system without knowing what results come from the projection, we never think of it as harnessing. In a natural system, such as in an oracle computation, scientists cannot make every step completely explicit; however, they know what happens if they throw artifacts (constructive computation) into the natural system. They indirectly but almost perfectly 'control' it (cf. 3.5 above).

5.4   Other Key Points

Just as every achievement brings us pleasure, so achievement in natural computing brings us pleasure. As for invention, researchers in the field of natural computing discover algorithms in nature, and their discovery probably results from their inventive devices. Discovery also brings us pleasure, and it makes us aware of complexity in nature.

Our goal here is not to decide whether natural computing is the genuine successor to fine arts or not, but to make clear that the idea of natural computing shares some key points with the original idea of fine arts. In the final section, we would like to suggest from the point of view of natural computing, especially from the point of view of 'harnessing,' what artworks we can evaluate as possible successors to the original idea of fine arts.

5.5   What Should We Evaluate?

If we look back to past artworks from the point of view of natural computing, can we find possible successors to fine arts?

Take Walter De Maria's *Lightning Field* (1971-77) for example, though it is one of the most famous artworks in the 20th century and it has never been overlooked. However, while this work has been interpreted in the context of conceptual and critical art (Godfrey 1998, Kastner 2010), we can reconsider it from the point of view of 'harness' because this work operates a larger system (the weather conditions which cause lightning) by using minimum artifacts (stakes made by stainless steel deliberately put on the field). Its significant difference from other land and environmental art is that the maker indirectly, but completely, controls what happens (cf. 3.5, 5.3).

From horticulture to this work, we can assume an imaginary lineage. In addition, as an extension of this lineage, we can expect the appearance of new successors to fine arts.

In this context, a series of activities by the following artists come to mind: Yasuhiro Suzuki and Rieko Suzuki (for example, Face Therapy), Junji Watanabe and his colleagues (for example, Saccade-based Displays), and Hiroya Tanaka and his colleagues (for example Fablab Japan). Even if each of them has already been highly appreciated in the field of facial massage, device arts, and media arts—and it is probably enough reason—, we would like to propose that we could think of their activities as successors to the original idea of fine arts. This is not only because their activities are realized through the idea of computation, which brought about the radical reorganization of arts in general (cf. 5.1), but also because they 'harness' human perception as a natural system with minimum artifacts (cf. 5.2) and furthermore they are completely responsible for the results of their works (cf. 3.5, 5.3).

6   Concluding Remarks: What Theory of Art Could Learn from Natural Computing

In this presentation, we looked back to a short history of arts. Through this reflection, we made clear that current critical artworks are not successors to original fine arts. Rather, we find that the idea of natural computing, especially the idea of 'harnessing,' shares some of the original key points of fine arts. Then, from the point of view of natural computing, we could reevaluate artworks and propose to make new lineages, for example, from horticulture in the 18th century through land art to the idea of harnessing in natural computing. We further suggest that we could expect new successors to extend this lineage from fine arts. We have learned these things from natural computing.

Acknowledgement. This work was supported by JSPS KAKANHI Grant Number 21520133.

References

Batteux, C.: Les Beaux Arts: Reduits a un meme principe. Slatkine reprints, Geneve (1969)

Beardsley, M.C.: Aesthetics: Problems in the Philosophy of Criticism (1958)

Daston, L., Peter, G.: Objectivity. The MIT Press (2007)

Diderot, d'Alembert.: Encyclopedie, ou Dictionnaire Raisonné des Sciences, des Artes et des Métiers, par une société de gens de lettres (1751)

Godfrey, T.: Conceptual Art. Phaidon (1998)

Kant, I.: Kritik der Urteilskraft (1790)

Kastner, J. (ed.): Land and Environmental Art. Phaidon (2010)

der Hypo-Kulturstiftung Muenchen, K.: Eine neue Kunst? Eine andere Natur! – Forografie und Malelei im 19. Jahrhundert, Schirmer/Mosel Produktion (2004)

Otabe, T.: Paradoxes of Art: Origin of Modern Aesthetics. The University of Tokyo Press (2001) (in Japanese)

Panofsky, E.: Idea: a concept in art theory. University of South Carolina Press (1968)

Shaftesbury, A.A.C.: The Moralists. In: Standard Edition: Complete Works, selected Letters and posthumous Writings, 2.1 Moral and Political Philosophy, frommann-holsboog (1987)

Shinya, K., Fujie, M., Suzuki, Y.: Possible? Modification of severe influenza diseases by harness. In: 6th International Workshop on Natural Computing, JSAI, p. 2 (2012)

Suzuki, Y.: Basics of natural computing. Kindai-Kagakusha (forthcoming)

Suzuki, Y.: What is natural computing?: From algorithm to harness. In: Symposium Bio-Aesthetica: Harness and Natural Computing, pp. 5–10 (2012) (in Japanese)

Wendler, R.: Das Spiel mit Modellen: Eine methodische Verwandtschaft künstlerischer Werk- und molekularbiologischer Erkenntnisprozesse. In: Reichle, I., Siegel, S., Spelten, A. (eds.) Visuelle Modelle, pp. 101–116. Wilhelm Fink Verlag (2008)

Study on the Use of Evolutionary Techniques for Inference in Gene Regulatory Networks

Leon Palafox[1], Nasimul Noman[2], and Hitoshi Iba[2]

[1] Department of Electric Engineering, University of Tokyo
[2] Department of Information, Science and Technology, University of Tokyo
{leon,noman,iba}@iba.t.u-tokyo.ac.jp

Abstract. Inference in Gene Regulatory Networks remains an important problem in Molecular Biology. Many models have been proposed to model the relationships within genes in a DNA chain. Many of these models use Evolutionary Techniques to find the best parameters of specific DNA motifs.

In this work, we compare the popular S-System using a powerful evolutionary technique, DPSO, and the novel Recursive Neural Network model, using clustered Population Based Incremental Learning (PBIL). We will use the SOS network for *E.coli* to do the comparison to finally show how they fare against other techniques in the area of Gene Regulatory Network (GRN) inference.

1 Introduction

Finding the correct interactions within genes' motifs is an important problem in Molecular Biology. There are many techniques to find the different interactions between genes in a Gene Regulatory Network (GRN). Most of these techniques have different strengths and weaknesses as we try to find interactions in larger networks.

In this work, we present a comparison between two techniques used to find the best parameters of a GRN. On one side, we use the S-System, an ODE modeling framework to find the correct interactions, which has been used by many groups in the area [1,2]. On the other hand, we use the novel Recursive Neural Network (RNN) architecture, recently proposed as a surrogate to model the interactions in a GRN [3].

To find the best parameters in both the S-System and the RNN, we use two different evolutionary techniques, we compare the effectives of random jumps of the Dissipative Particle Swarm Optimization (PSO)[4] against clustering the candidate solutions to enrich the search space in Clustered Population Based Incremental Learning (PBIL), which we call K-Means PBIL (KPBIL). We use PSO to solve the S-System and KPBIL to solve the RNN.

The paper is organized as follows; first, we present a brief introduction to both techniques, we show how to use the S-System and RNN to do reverse engineering of GRN. Then, we introduce the two evolutionary techniques we used, Dissipative PSO and KPBIL. Then, we describe how to do inference for the parameters over small sets and finally use both algorithms to do inference in the real SOS network for *E.coli*.

Y. Suzuki and T. Nakagaki (Eds.): WSH 2011 and IWNC 2012, PICT 6, pp. 82–92, 2013.

1.1   Related Work

Different researchers have attacked the problem of gene network inference using different techniques. These techniques have strengths and weaknesses, of which we will describe some.

One of the most intuitive approaches is the use of graphical models to represent the genes and its connections, Akutsu [5] used a boolean network to represent the connections between genes as logical connections. This model, albeit powerful is limited in its capability of representing the true state of genes over time, and was restricted to tree-like structures. Murphy [6], however, showed that these models could be best represented as dynamic Bayesian networks, which allow for structures other than trees to represent data along time. Bayesian Networks, however, have difficulties to deal with the data when it is represented in loops, like in real GRN.

Other family of widely used models are the differential equations based models (ODE), these models use non linear differential equations to represent dynamic data. The S-System, an ODE modeling framework, is extensively used by many researchers in the area of evolutionary computation [2,7,8]. ODE modeling frameworks have many advantages compared to graphical representations, since they allow for a richer representation of the data, however, solving a set of differential equations is computationally more expensive than inferring probabilistic graphical models, making them unfeasible to work with larger networks.

Vohradský[9], proposed using RNNs to model gene networks. Furthermore, Xu et al [3] successfully applied Particle Swarm Optimization (PSO) to find the parameters of the RNN. Different works [10,11] have used evolutionary computation techniques to solve Neural Network's architectures with good results. This model has some of the advantages of the graphical models, like the scalability, and the advantage of richness of expression that the ODE modeling frameworks offer.

2   GRN Modeling

To model the GRN, we use training data in the form of microarray data, this data is the expression of different genes over a period of time. We use two different models and compare them, the S-System and the RNN model.

2.1   S-System as a Model for Biological Networks

The S-System, first proposed by Savageau [12], provides a mathematical framework to represent and analyze biological systems. It represents a network as a set of differential equations having the form:

$$\frac{dX_i}{dt} = \alpha_i \prod_{j=1}^{N} X_j^{g_{ij}} - \beta_i \prod_{j=1}^{N} X_j^{h_{ij}} \tag{1}$$

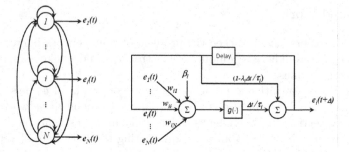

Fig. 1. Left: RNN's graphical model with the observed data, Right:RNN scheme wit the addition of delay terms and the output for a single gene

where X_i is the expression level of the *ith* gene of the network, N is the number of genes in the network, $\alpha_i, \beta_i \in \mathbb{R}_+^N$ are *rate constants* and $g_{ij}, h_{ij} \in \mathbb{R}^N$ are *kinetic orders*. It is worth mentioning that the *kinetic orders* g_{ij} and h_{ij} affect the synthesis and degradation of X_i due to X_j.

2.2   RNN as a Model for GRN

Since the inputs for a classic Neural Network are taken iid from the training set, NNs are not suited to model temporal data. The RNN model, however, is a closed loop NN with a delay variable suitable to model dynamic systems (Fig. 1).

The model connects the output of each neuron in the output layer of the RNN to each of the neurons in the input layer via a delay parameter in Fig 1. Vohradský [9] modeled the gene's regulations with this architecture, assuming that each of the neurons in the output unit $(e_i(t + \Delta t))$ is a gene, and the neurons in the input units$(e_{i,...,N}(t))$ are the same genes, thus every gene interacts with one another.

The mathematical model of an RNN resembles a standard NN with additional variables for the feedback loop. The discretized version has the following form:

$$e_i(t + \Delta t) = \frac{\Delta t}{\tau_i} \times f\left(\sum_{j=1}^{N} w_{ij}e_j(t) + \beta_i\right) + \left(1 - \frac{\Delta t}{\tau_i}\right)e_i(t) \qquad (2)$$

where, $f()$ is a nonlinear function that acts as a classification function, we use the sigmoid function $f(z) = 1/(1 + e^{-z})$. The values w_{ij} are the connecting weights of the network, which represent the connections between gene i and j. The variable e_j represents the expression level for the gene, which is the data we receive from the microarray experiments. And finally, β is the bias parameter of the network.

Forward evaluation evaluates an initial input using the set of weights $W_l = \{w_{ij}\}$ in Eq. 2 and obtains a new time series $e_i(t)$.

Xu et al [3] used PSO to find the weights of an RNN focused on the problem of GRN inference. Here, we used an improved version of PBIL to find the weights of the RNN.

2.3 Estimation Criteria and Regularization

To find the best parameters for the network Tominaga [13] standardized the use of the Mean Squared Error (MSE) evaluation to measure a candidate's fitness in the S-System. They defined the fitness function as:

$$f = \sum_{t=1}^{T} \sum_{i=1}^{N} \left(\frac{X_{i,cal}(t) - X_{i,exp}(t)}{X_{i,exp}(t)} \right)^2 \tag{3}$$

where T represents the number of time samples in the experimental data, N is the number of genes, and cal, exp refer to the calculated and experimental values of the gene expression's data, respectively. In the case of the S-System, X_{cal} represents the solution of Eq. 1 and for the RNN, it is the output of Eq. 2.

3 Optimization Methods

To find the values that best fit equation and minimize equation 3 using the RNN and S-System models, we use 2 different optimization algorithms from the area of evolutionary computation, which are Dissipative PSO and Population Based Incremental Learning (PBIL). These two methods are described in the following sections and it has to be noted that the described equations are for the general case of its implementation.

3.1 Dissipative PSO

Particle Swarm Optimization (PSO) is an algorithm widely used in different applications. It has been used to analyze human activities [14], optimize power networks [15] and for scheduling problems [16].

In PSO a swarm is composed of particles, each of which can record its best position, and the swarm's best position. Each particle also has a velocity, which updates to grow closer to the best positioned particle in the swarm at each iteration. We define each particle p_i at time $t \in (0, IT-1)$ as $p_i(t) \in \mathbb{R}^N$, where IT is the maximum number of iterations and N is the feature space's dimension. The variables that control each particle at time t are its position $p_i(t)$, and its velocity $v_i(t)$. Each particle's velocity and position will change according to:

$$v_i(t+1) = w \cdot v_i(t) + C_1 \cdot \varphi_1 \cdot (P_{li} - p_i(t)) \tag{4}$$
$$+ C_2 \cdot \varphi_2 \cdot (P_G - p_i(t))$$
$$p_i(t+1) = p_i(t) + v_i(t+1) \tag{5}$$

where P_{li} is the best position for particle i, P_G is the best position the swarm has obtained and w is the inertia factor, which controls the speed at which the particles adapt. C_1 and C_2 are random variables that control the dependence on the global closeness and the φ_1, φ_2 factors are manually set variables to control the swarm's convergence.

Classic PSO, however, often reaches a local minimum as its final solution. Xiao et al. [4] proposed a variation to the classic PSO, adding a dissipative property to the particles. They defined dissipation parameters that restart the system at random iterations. The dissipative equations, evaluated at each iteration, are:

$$IF(rand() < c_v) => v_i = rand() * v_{max} \tag{6}$$

$$IF(rand() < c_l) => x_{id} = Random(l_d, u_d) \tag{7}$$

where c_v and c_l are numbers between 0 and 1. Setting small numbers to these variables result in few restarts, allowing each new iteration to reach a new optimum. Variables v_{max}, l_d, u_d are the particle's velocity limit, and the lower and upper bound for the search space respectively. The random variables $rand()$ and $Random(I_d, v_d)$ correspond to an uniform random sample and to a sample from the interval given by the lower and the upper bound.

For the S-System, we will define each particle of the swarm with a variable vector p_i, composed of the parameters $\theta_i \in \{g_{ij}, h_{ij}, \alpha_i, \beta_i | i, j \in 1 \dots N\}$.

3.2 PBIL and Clustering

Population Based Incremental Learning (PBIL)[17] is an optimization technique that finds the best candidates of a function by inferring a probability distribution from each of the dimensions in the feature set. This creates N probability distributions, where N is the problem's dimension.

The algorithm chooses the best candidates using the fitness function 3, and then sets a threshold that selects only the best candidates to infer a new probability distribution. New samples will be in turn be taken from this distribution, and a new fitness process will be done. This sampling-fitting process is done recursively until the variations are so low that we will have reached a minimum for the fitness function.

Since the search space is non-convex, if the problem is modeled using standard Gaussians distributions, we are modeling a multimodal problem with an unimodal distribution. Thus conveying local minimum instead of global minimum. In his work, Emmendorfer[18] showed that a mixture of Gaussians, as a multimodal distribution, is a good alternative to model non-convex search spaces.

There are different ways to create a mixture of Gaussians, like doing expectation maximization (EM), or a more naive clustering technique like K-means. K-means are a relaxation of a mixture of Gaussian distributions with symmetric variances, and is a faster procedure than EM for estimation of a mixture of Gaussians.

Using K-means we model the candidates' search space as a mixture of K Gaussian distributions, to have at a set of $N \times K$ clusters modeling the best candidates of the problem for each of the N dimensions.

4   Algorithm to Infer the GRN

The inference algorithm does the following steps:

1. Generate N candidate solutions $X_i \in R^N$ from an uniform distribution $X_i \sim U(X_{min}, X_{max})$.
2. Using the candidates, do the forward evaluation of the RNN or solve the S-System given in equation 2 and 1, respectively, and obtain a set of time series TS per each candidate.
3. Using the cost function (Eq. 3), rank the candidates and choose the best M ones.
4. Update the candidate solutions:
   - (PSO) Using all the candidates, update their values using Eq. 4
   - (PBIL)Using the M candidates, generate $N \times K$ Gaussian distributions using K-means and variance equal to 1.
   - (PBIL)Generate N new candidates from the inferred Gaussian distributions.
5. Go to step 2.

This recursive process is repeated until convergence conditions, which are set by us, are met.

5   Experiments

We tested both models, the S-System with PSO and the RNN with KPBIL, with the popular SOS network for $E.coli$ [19]. We ran both of them, and counted the total number of true positives and true negatives to calculate measurement variables like Recall, Precision and F-Score.

To test the effect of the variables in the PBIL, we changed the population size, from 100 to 500 candidates, as well as the cluster number in the K-means, from 1 to 5, to test whether multiple clusters do better than a single probability density. For the PBIL implementation, we did 2000 iterations per run, with each iteration lasting at most 5 minutes for the architectures with 500 candidates.

In PSO, we evaluated the effect of the population as well, and variated it from 20 to 320. We did 5000 iterations per run, with each of them evolving 20 particles. Each of the iterations took 3.2 hours to reach a steady state.

To evaluate the variables, we used the Recall, Precision, True Negative Rate (1-Specificity(S_p)) and F-Score using the following equations:

$$Recall = \frac{TP}{TP+FN} \qquad\qquad Precision = \frac{TP}{TP + FP}$$

$$S_p = \frac{TN}{TN+FP} \qquad\qquad F - Score = 2\frac{Precision * Recall}{Precision + Recall}$$

where TP, FP, TN and FN stand for True and False positive and negative, respectively. In our approach, a true positive is assigned when a connection between 2 genes is truth and a true negative, when the absence of the connection is true.

5.1   Real SOS Network

The SOS network[19] for *E.coli*, published by the Uri Alon group (Fig. 2), is a benchmark for testing genetic regulation inference.

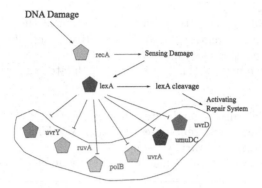

Fig. 2. Graphical Representation of the SOS Net, where the lexA represses every gene

In their experiments, 8 genes are expressed (uvrD, lexA, umuD, recA, uvrA, uvrY, ruvA and polB). They irradiate the DNA with UV light, which affects some genes, after that, the network will repair itself. They did four experiments for different light intensities. Each experiment had 50 time steps spaced by 6 minutes. Many researchers, however, usually choose to infer only 6 of the 8 genes, since two of them have marginal activity in comparison with the rest of the genes in the network. For the sake of comparison, we have worked also exclusively with 6 genes.

5.2   Comparative Analysis on the Population Size and Number of Clusters

We plotted the Recall for different population sizes and number of clusters in Figs. 3a and 3b.

Fig. 3a shows that for few clusters, the recall is good, and generally, large populations present marginally better results than small populations. This is because the K-means method does not require many particles to create reliable clusters, furthermore, clustered approaches have better results than non-clustered approaches. For this problem, then, using more than 3 clusters seems to dampen the search, and a few clusters —2 or 3— is the best alternative. We can see as well that the unclustered option —1 cluster— has worst results than its alternatives,

Fig. 3b shows that for large populations the recall of the DPSO is reduced by a small factor, this is because we let them run for the same time. Intuitively, large populations will spend more resources than small populations, so it takes longer to converge to a good solution. The results, however, show that the DSPO algorithm is capable of obtaining good results for a small number of particles.

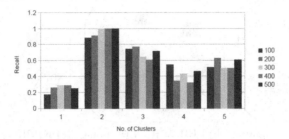

(a) Recall for Different Clusters and Population for the
KPBIL. Evaluation for 100 to 500 possible candidates.

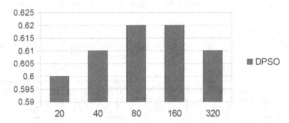

(b) Recall values for different population sizes using
DPSO

Fig. 3. Recall mean and standard deviation for the SOS Network

In Fig. 4 we show a qualitative solution to the problem of finding the right
connections in the SOS Network. Fig. 3b shows the result of using DPSO, which
has many false positives, related with the low recall that we presented before.

In Fig. 4b we see the inferred network using a traditional PBIL, without any
clustering to infer multiple probabilities distributions. While the performance
was better than the DPSO, it still finds many false positives, specially on self
regulations.

Finally, Fig. 4c presents the results using the KPBIL approach, which has
the most promising results of them three. The network however becomes sparse,
losing some important connections.

To compare our results, we compiled results from other papers working with
the SOS Net. Table 1 shows this compilation, here as well we have a quantitative
comparison of our approach.

Table 1 shows that KPBIL does inference better than other approaches that
use both evolutionary and non-evolutionary techniques. KPBIL is much faster
than similar approaches, specially comparing with the state of the art, which is
the S-Tree, which takes 35 hours to do the whole inference. DPSO presents one
of the best sensitivities of the group, this is because the S-System iw one of the
most precise models for these kind of problems, while the RNN model is just an
approximation that is used for the sake of speed but at the expense of precision.

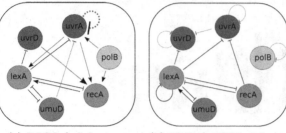
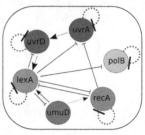

(a) DPSO & S-System   (b) PBIL & RNN with one   (c) KPBIL & RNN with 3
                              cluster   clusters

Fig. 4. Solutions for different combinations of PBIL and DPSO, the straight lines are True connections while the red dotted lines are spurious connections

Table 1. Specificity, Sensitivity and F-Score for different models of the *E.coli* SOS Network

	#Regs	#TP	#FP	#TN	#FN	Sensitivity	Specificity	F-Score	Time[h]
Bayesian Network[20]	6	4	2	18	3	0.571	0.900	0.615	0.01
S-Tree[21]	7	6	1	19	1	0.857	0.950	0.857	35
LTV[22]	13	7	6	14	0	1.000	0.700	0.7	0.1
DE[1]	8	5	3	17	2	0.714	0.850	0.667	0.3
KPBIL	11	7	4	13	3	0.7	0.765	0.67	0.05
DPSO	10	7	3	17	0	1	0.68	0.60	3.5

6   Conclusions

We have presented a comparison between two methods to do inference in Gene Regulatory Networks, a Recursive Neural Network with PBIL and the S-System using DPSO. We have done comparison and analysis of both methods using the SOS network for *E.coli*, which is a benchmark for small network inference. We presented in the results that both RNN and PBIL present promising results for the inference problems, it had both better results and faster inference rate, which are two of the most important desiderata in molecular biology.

For future work, we will attempt the use of our model for larger networks, modeling larger networks remains a challenge for this kind of approaches, since the complexity increases vastly with each new gene we add to the system.

References

1. Noman, N., Iba, H.: Inferring gene regulatory networks using differential evolution with local search heuristics. IEEE/ACM Transactions on Computational Biology and Bioinformatics 4(4), 634–647 (2007)
2. Kikuchi, S., Tominaga, D., Arita, M., Takahashi, K., Tomita, M.: Dynamic modeling of genetic networks using genetic algorithm and S-system. Bioinformatics 19(5), 643 (2003)
3. Xu, R., Donald Wunsch, I.I., Frank, R.: Inference of genetic regulatory networks with recurrent neural network models using particle swarm optimization. IEEE/ACM Transactions on Computational Biology and Bioinformatics, 681–692 (2007)
4. Xie, X.F., Zhang, W.J., Yang, Z.L.: Dissipative particle swarm optimization. In: Proceedings of the 2002 Congress on Evolutionary Computation, CEC 2002, vol. 2, pp. 1456–1461. IEEE (2002)
5. Akutsu, T., Miyano, S., Kuhara, S.: Identification of genetic networks from a small number of gene expression patterns under the Boolean network model. In: Pacific Symposium on Biocomputing, vol. 4, pp. 17–28. Citeseer (1999)
6. Murphy, K., Mian, S.: Modelling gene expression data using dynamic Bayesian networks. Graphical Models, 12 (1999)
7. Noman, N., Iba, H.: Inference of gene regulatory networks using s-system and differential evolution. In: Proceedings of the 2005 Conference on Genetic and Evolutionary Computation, Washington, DC, p. 439. Citeseer (2005)
8. Kimura, S., Ide, K., Kashihara, A., Kano, M., Hatakeyama, M., Masui, R., Nakagawa, N., Yokoyama, S., Kuramitsu, S., Konagaya, A.: Inference of S-system models of genetic networks using a cooperative coevolutionary algorithm. Bioinformatics 21(7), 1154 (2005)
9. Vohradský, J.: Neural network model of gene expression. The FASEB Journal: official publication of the Federation of American Societies for Experimental Biology 15(3), 846–854 (2001)
10. Pettersson, F., Biswas, A., Sen, P.K., Saxén, H., Chakraborti, N.: Analyzing Leaching Data for Low-Grade Manganese Ore Using Neural Nets and Multiobjective Genetic Algorithms. Materials and Manufacturing Processes 24(3), 320–330 (2009)
11. Zamparelli, M.: Genetically Trained Cellular Neural Networks. Neural Networks: The Official Journal of the International Neural Network Society 10(6), 1143–1151 (1997)
12. Savageau, M.A.: Biochemical systems analysis+*:: I. Some mathematical properties of the rate law for the component enzymatic reactions. Journal of Theoretical Biology 25(3), 365–369 (1969)
13. Tominaga, D., Koga, N., Okamoto, M.: Efficient numerical optimization algorithm based on genetic algorithm for inverse problem. In: Proceedings of the Genetic and Evolutionary Computation Conference, vol. 251, p. 258 (2000)
14. Palafox, L., Hashimoto, H.: 4W1H and Particle Swarm Optimization for Human Activity Recognition. Journal of Advanced Computational Intelligence and Intelligent Informatics 15(7), 793–799 (2011)
15. AlRashidi, M., El-Hawary, M.: A survey of particle swarm optimization applications in electric power systems. IEEE Transactions on Evolutionary Computation 13(4), 913–918 (2009)
16. Liao, C., Luarn, P.: A discrete version of particle swarm optimization for flowshop scheduling problems. Computers & Operations Research 34(10), 3099–3111 (2007)

17. Sebag, M., Ducoulombier, A.: Extending population-based incremental learning to continuous search spaces. In: Eiben, A.E., Bäck, T., Schoenauer, M., Schwefel, H.-P. (eds.) PPSN 1998. LNCS, vol. 1498, pp. 418–427. Springer, Heidelberg (1998)
18. Emmendorfer, L., Pozo, A.: Effective Linkage Learning Using Low-Order Statistics and Clustering. IEEE Transactions on Evolutionary Computation 13(6), 1233–1246 (2009)
19. Ronen, M., Rosenberg, R., Shraiman, B.I., Alon, U.: Assigning numbers to the arrows: parameterizing a gene regulation network by using accurate expression kinetics. Proceedings of the National Academy of Sciences 99, 10555 (2002)
20. Perrin, B.E., Ralaivola, L., Mazurie, A., Bottani, S., Mallet, J., D'Alche-Buc, F.: Gene networks inference using dynamic Bayesian networks. Bioinformatics 19(suppl. 2), ii138–ii148 (2003)
21. Cho, D.Y., Cho, K.H., Zhang, B.T.: Identification of biochemical networks by S-tree based genetic programming. Bioinformatics 22, 1631–1640 (2006)
22. Kabir, M., Noman, N., Iba, H.: Reverse engineering gene regulatory network from microarray data using linear time-variant model. BMC Bioinformatics 11(suppl. 1), S56 (2010)

Reconstruction of Gene Regulatory Networks from Gene Expression Data Using Decoupled Recurrent Neural Network Model

Nasimul Noman, Leon Palafox, and Hitoshi Iba

Graduate School of Information Science and Technology,
University of Tokyo, Tokyo 113-8656, Japan
{noman,leon,iba}@iba.t.u-tokyo.ac.jp

Abstract. In this work we used the decoupled version of the recurrent neural network (RNN) model for gene network inference from gene expression data. In the decoupled version, the global problem of estimating the full set of parameters for the complete network is divided into several sub-problems each of which corresponds to estimating the parameters associated with a single gene. Thus, the decoupling of the model decreases the problem dimensionality and makes the reconstruction of larger networks more feasible from the point of algorithmic perspective. We applied a well established evolutionary algorithm called differential evolution for inferring the underlying network structure as well as the regulatory parameters. We investigated the effectiveness of the reconstruction mechanism in analyzing the gene expression data collected from both synthetic and real gene networks. The proposed method was successful in inferring important gene interactions from expression profiles.

Keywords: Recurrent Neural Network model, gene network reconstruction, decoupled RNN, differential evolution.

1 Introduction

In recent years, with the advent of various gene expression assaying techniques, the study of the relationship among genes has been highlighted extensively. Gene expression data, whether in time-course format or steady state format, provides an opportunity to observe the interaction among thousands of genes simultaneously. Given that sufficient amount of gene expression data is available, in principle it is possible to derive the detailed quantitative model of the network that adequately represents the dynamics of the underlying system [1].

Nevertheless, several practical issues such as small sample size compared to the number of genes, the presence of biological noise and experimental noise, inadequate knowledge and representation of the complex dynamics and nonlinear relationship among the genes, the problem dimension, makes the adequate reconstruction of the network a very challenging task [2]. Several techniques have been proposed in the field of computational intelligence for algorithm based reconstruction of gene regulatory networks (GRN) that help biologists to form new hypotheses about biological systems and to design new experiments [2–4].

Y. Suzuki and T. Nakagaki (Eds.): WSH 2011 and IWNC 2012, PICT 6, pp. 93–103, 2013.

In order to apply a computational approach for inferring GRN from experimental data, a mathematical model is necessary to formalize the process of gene regulation. The modeling endeavor for GRN, started long ago, produced a large variety of models over the last couple of decades. The modeling of GRN has diverged in many directions such as: discrete versus continuous, linear versus non-linear, deterministic versus stochastic, graphical versus non-graphical, synchronous versus asynchronous etc. All of these modeling paradigms have their strength and weakness in terms of representation accuracy, computational feasibility, noise proneness and data requirement [5].

In this work we have used the recurrent neural network (RNN) model [6] along with a natural computational method to extract regulatory interactions among genes from gene expression data sets. The canonical RNNs are neural networks with delayed feedbacks. With the network of nonlinear processing elements, RNN can adequately capture the nonlinear and dynamic interaction among genes [7]. Many of the reconstruction approaches applied to infer GRN using RNN belongs to the field of natural computation. Some of the researchers used genetic algorithms (GA) for reconstructing the target network using the single layer RNN [6, 8] and the multilayer RNN [9] architecture. A couple of swarm intelligence approaches have been proposed in which swarm algorithms (particle swarm optimization and ant colony optimization) have been applied for estimating network structure and parameters [7, 10]. Evolutionary algorithms other than GA have also been used for estimating RNN parameters for GRNs [11]. Some hybrids of natural computations with other approaches have been used also for reverse engineering GRNs using RNN formalism [2, 12].

The RNN model offers a good compromise between the biological proximity and mathematical flexibility for representing GRNs. However, inference of GRN using RNN requires the estimation of $N(N+3)$ parameters, where N is the number of genes in the network. Generally, with the increase of the dimension, the problem complexity increases rapidly and locating the global optimum solution becomes difficult for the search algorithm. Therefore, in order to deal with the challenges of high-dimensionality, with increasing genes in the network, here we use a decoupled form of the RNN model for inferring GRN. Other than reducing the problem dimension, such decoupling facilitates the design of parallel algorithms for GRN inference. In this work we have used a natural computational approach called differential evolution (DE), belonging to the group of evolutionary algorithm, for identifying the regulatory interactions form expression profiles using the decoupled form of RNN. We tested the proposed method using artificial gene regulatory networks of different dimensions and a real network. Experiments showed that the proposed approach can provide a good estimate of the structure of genetic networks.

The rest of the paper is organized as follows. The next section describes the decoupled RNN model. In section 3, we present the DE algorithm for inferring RNN model based gene networks. The fourth section reports the experiments with the results to verify the effectiveness of the proposed method. In section 5 we conclude the paper with some general discussions.

2   Recurrent Neural Network (RNN) Model

The recurrent neural network (RNN) model formulates the genetic interactions in terms of a neural network in which the nodes correspond to genes and the connections correspond to regulatory interactions among genes [6, 13]. In canonical RNN model the interactions among genes is represented in terms of a tightly coupled system given by

$$\frac{de_i}{dt} = \frac{1}{\tau_i} \left(g \left(\sum_{j=1}^{N} w_{ij} e_j + \beta_i \right) - \lambda_i e_i \right) \tag{1}$$

where e_i represents the gene expression level for the i-th gene ($i \leq N$, N is the total number of genes in the network). w_{ij} represents the type and strength of the regulatory interaction from j-th gene towards i-th gene. The positive (negative) value of w_{ij} represents activation (repression) control of gene-j on gene-i. $w_{ij} = 0$ means gene-j has no regulatory control on gene-i. β_i represents the basal expression level and λ_i denotes the decay rate parameter of the i-th gene. The function $g(\cdot)$ introduces non-linearity to the model which is often given by the sigmoid function. In the canonical form the RNN model for GRN can be described by the following set of $N(N+3)$ parameters $\Omega = \{w_{ij}, \beta_i, \lambda_i, \tau_i\}$ where $i, j = 1, 2, \cdots, N$.

In this work we estimated the regulatory interactions towards a particular gene at a time, independent of interactivity on other genes. In other words, we have divided the $N(N + 3)$ dimensional problem into N sub-problems of size $(N + 3)$ and solved each separately. In sub-problem i ($i = 1, 2, \cdots, N$) we estimated the model parameters $\Omega_i = \{w_{ij}, \beta_i, \lambda_i, \tau_i\}$ ($j = 1, 2, \cdots, N$). Then we accumulate all the learned parameters to build the complete network model. Similar procedure for learning the interactions separately has been applied with other neural network models [14] or with some other GRN models [15, 16]. In addition to reducing the problem dimension, this decoupling procedure makes the parallel solution of each sub-problem possible.

3   Reverse Engineering Algorithm

Here we used a natural computational approach for reverse engineering GRN modeled by decoupled RNN. We employed differential evolution (DE) [17] for searching the optimal model parameters that can reproduce the target time courses of genes. DE is a new generation EA proven to be very successful in solving different complex problems arising in different domains. It has also been very effective in reverse engineering GRN using the canonical RNN model [11, 18]. Hence, we chose DE for identifying the regulatory interactions among genes using the decoupled RNN formalism.

Like most of the EAs, DE starts with a population of random solution where each individual of the population encodes a candidate solution for the problem under consideration. Here we apply a separate instance of DE for estimating the parameters of each target gene in the network. In other words, in sub-problem i, every individual of DE represents the parameters for gene-i that is $\Omega_i = \{w_{ij}, \beta_i, \lambda_i, \tau_i\}$ $(j = 1, 2, \cdots, N)$. After random initialization, the fitness of the individual is calculated using the fitness function described later. Then for each individual x_G^i, $i = 1, \cdots, P$ in the current generation, G, a mutated individual is generated by the following *mutation* operation

$$y_G^i = x_G^j + F(x_G^k - x_G^l) \tag{2}$$

where x_G^j, x_G^k and x_G^l are random individuals selected from generation G such that j, k and $l \epsilon \{1, \cdots, P\}$ and $i \neq j \neq k \neq l$; P is the number of individuals in G and F is called the *scaling factor* – a parameter of DE.

Afterwards, the mutated individual y_G^i participates in a *crossover* operation with the current population member x_G^i and generates the *offspring* y_{G+1}^i. In the *crossover* operation, the genes (parameters) of the offspring y_{G+1}^i are randomly inherited from x_G^i or y_G^i determined by a parameter called *crossover factor* CF, i.e. if $r \leq CF$ (where r is a uniform random number in $[0, 1]$) then it is inherited from x_G^i otherwise from y_G^i. Finally, the offspring is evaluated and replaces its parent x_G^i in next generation if its fitness is the same or better than its parent. This is the *replacement* process for producing new generation. And this process is iterated generation after generation until a satisfactory solution is found or a maximum number of generations (G_{max}) have elapsed.

Because of the flexibility of the model, the search space contains many local optimum that traps the search algorithm and the global optimum remains undiscovered. In order to help the algorithm to get out of a local optimum we embedded a random restart strategy in DE that randomly reinitializes all the individuals except the elite one, if the difference between the best fitness (f_{best}) and the worst fitness (f_{worst}) of the current generation falls below a threshold $(\delta \cdot f_{best})$. After the random restart, the algorithm proceeds in its regular mode. When the optimization of one instance of DE finishes, we receive the set of parameters for one gene. Repeating this process for all genes and compiling them together we get the complete set of parameters for the whole network.

3.1   Fitness Evaluation Criteria

We need some assessment mechanism for evaluating the alternate GRN models we come across in course of the evolutionary process. The most commonly used model evaluation process is the quantitative difference between the response generated by the candidate model and the experimentally collected response. This evaluation process calculates the model fitness using a function called mean squared error (MSE). The reverse engineering of GRN, like other dynamic systems, can be done with higher accuracy if multiple time series for the same gene could be used. Since we are estimating the parameters of each gene separately, the fitness evaluation process takes the time courses of a particular gene in consideration. Using M sets of time dynamics, the MSE based fitness function for the sub-problem i, corresponding to gene-i, is given by

$$f(\Omega_i) = \frac{1}{TM} \sum_{k=1}^{M} \sum_{t=1}^{T} \left(e_{k,i}^{cal}(t) - e_{k,i}^{exp}(t) \right)^2 \tag{3}$$

where $e_{k,i}^{exp}(t)$ and $e_{k,i}^{cal}(t)$ represent the expression levels of i-th gene in the k-th set of time courses at time t in experimental and simulated data respectively.

Generally, very few genes or proteins regulate the expression level of a specific gene [19]. But the general model of RNN considers every possible interaction from each gene. Because of the model flexibility, if we allow all possible regulations then the search algorithm gets stuck to some local minimum that can generate the time course very closely. One effective way of recovering the target skeletal structure is to penalize the fitness score in proportional to the network complexity [20, 21]. Here we use a penalty term similar to that used in [22]. The penalized fitness function that was used for evaluating the models are given by

$$f(\Omega_i) = \frac{1}{TM} \sum_{k=1}^{M} \sum_{t=1}^{T} \left(e_{k,i}^{cal}(t) - e_{k,i}^{exp}(t) \right)^2 + c \sum_{j=1}^{N-I} (\widehat{w_{ij}}) \tag{4}$$

where $\widehat{w_{ij}}$ are the weights of interactions towards gene-i sorted in ascending order of their magnitude. I indicates a limit on the maximum allowed regulations for a gene. If the number of interactions exceeds this limit then the pruning term will penalize the fitness function. c represents the penalty constant.

4   Experimental Results

In this work, the suitability of the GRN inference using the decoupled RNN model was primarily validated using synthetic networks since the actual structure and parameter values are unknown for real networks. Two different networks of different sizes and architectures were used for this purpose. The reconstruction experiments were carried out under the ideal noise-free condition and with simulated noise corrupted gene expression data. We also attempted to reconstruct the SOS DNA repair network of *Echericha coli* using the proposed method.

4.1   Artificial Network Inference

In our first experiment with *in silico* networks, we investigated whether it is possible to infer the regulatory interactions and correct parameter values for a target gene network using the decoupled form of RNN model. In the reconstruction experiment we used a small scale network that has been studied by others in canonical RNN model [6, 10, 11]. The parameters for the RNN model for this four gene network, called NET1 hereafter, is shown in Table 1. In NET1 $\lambda_i = 1$ used for all genes as done in other work [6, 10, 11].

The artificial gene expression data was generated by simulating the canonical model of RNN in Table 1. The initial gene expression level was selected randomly. We generated $M = 10$ sets of gene expression profiles for NET1 where each time

Table 1. RNN model of the target synthetic network NET1

w_{ij}	1	2	3	4	β_i	τ_i
1	20.0	-20.0	0.0	0.0	0.0	10.0
2	15.0	-10.0	0.0	0.0	-5.0	5.0
3	0.0	-8.0	12.0	0.0	0.0	5.0
4	0.0	0.0	8.0	-12.0	0.0	5.0

Table 2. Inferred RNN model for NET1 from 5% noisy data

w_{ij}	1	2	3	4	β_i	τ_i
1	49.99	-20.14	-7.79	0.00	-3.33	10.04
2	18.15	-12.28	0.51	0.00	-6.21	5.19
3	0.00	-8.19	11.46	-0.77	0.48	4.25
4	0.00	-0.29	6.51	-9.20	-0.27	4.48

courses contains $T = 50$ time samples. In order to simulate the noise experienced in the real gene expression data we generated expression profiles adding 5% Gaussian noise. Then we tried to reverse engineer the target network from both the noise-free data and noise-corrupted data.

In our experiments, we inferred the regulators of gene-i $(i = 1, \cdots, N)$ under the same experimental condition. For every sub-problem the algorithmic setup was as follows: $F = 0.5$, $CF = 0.9$, $P = 100$, $G_{max} = 10000$, $\delta = 1 \times 10^{-3}$, $c = 10$ and $I = 4$. The setting for DE parameters $(F, CF, \text{ and } P)$ is very typical [23] and other parameters were chosen based on the setting used in [22] or empirically. The search ranges for RNN parameters were as follows: $w_{ij} \in [-30.0, 30.0]$, $\beta_i \in [-10.0, 10.0]$, $\tau_i \in [0.0, 20.0]$. We did not include λ_i in our search as it was fixed in the target model. We implemented the algorithm in Java and experiments were run in a Intel® Core™$i7$ CPU 2.67 GHz computer with 8GB RAM. Each experiment was repeated 10 times to confirm the reliability of the stochastic search algorithm.

In the reconstruction experiments from noise-free gene expression data we could precisely estimate the network structure and the parameter values. In almost every optimization run the fitness score for the models reached to zero or very close to zero $(< 1 \times 10^{-15})$ and the estimated parameters were exactly the same as the target. Although it was a very simple and small network, these experiments verify that if sufficient expression data is given and the dynamics are free from noise, then it is possible to estimate the network structure and kinetics using the decoupled form of RNN model.

We also analyzed the performance of the reconstruction algorithm in inferring NET1 from noisy expression data. The experimental condition was exactly the same as before except $I = 3$ was used. Table 2 shows the estimated network structure and parameter values achieved in a sample run. From Table 2 it is evident that even in presence of noise all the regulatory interactions among the genes were identified correctly. However, the estimated parameter values for the

Table 3. Summary of the results of NET1 reconstruction from 5% noisy data

Measure	5% noisy data
Sensitivity	1.000
Specificity	0.500
Accuracy	0.750
MCC	0.577

Table 4. RNN model of the target synthetic network NET2

$w_{i,j}$	$w_{1,14} = -15$, $w_{5,1} = 10$, $w_{6,1} = -20$, $w_{7,2} = 15$, $w_{7,3} = 10$, $w_{8,4} = 20$,
	$w_{9,5} = -20$, $g_{9,6} = 10$, $g_{9,17} = 10$, $w_{10,7} = -10$, $w_{11,4} = -15$, $g_{11,7} = 15$,
	$w_{11,22} = -15$, $w_{12,23} = 10$, $w_{13,8} = 20$, $w_{14,9} = 15$, $w_{15,10} = -10$, $w_{16,11} = 15$,
	$w_{16,12} = -15$, $w_{17,13} = -20$, $w_{19,14} = -15$, $w_{20,15} = 10$, $w_{21,16} = -20$, $w_{23,17} = -10$
	$w_{24,15} = -15$, $w_{24,18} = -20$, $w_{24,19} = 15$, $w_{25,20} = -10$, $w_{26,21} = 20$, $w_{26,28} = 20$,
	$w_{27,24} = -15$, $w_{27,25} = 10$, $w_{27,30} = 15$, $w_{28,25} = -15$, $w_{29,26} = 10$, $w_{30,27} = 15$,
	other $w_{i,j} = 0.0$
β_i	$\beta_i = 5$ for $i = \{2, 5, 6, 10, 16, 24, 28\}$ $\beta_i = -5$ for $i = \{15, 17, 27\}$ otherwise $\beta_i = 0$
τ_i	$\tau_i = 10$ for $i = \{1, \cdots, 30\}$
λ_i	$\lambda_i = 1$ for $i = \{1, \cdots, 30\}$

network kinetics were not very precise. Additionally, some false positive regulations were predicted. Nevertheless, if we consider the magnitude of these false positives then it is obvious that those were pretty small compared to real regulations. The summary of the prediction in terms of sensitivity, specificity, accuracy and Mathew's correlation coefficient (MCC) is presented in Table 3. Here, we used the standard definitions for these measurements based on positive/negative value of w_{ij}. These results show that the prediction had a full 1.00 sensitivity and a MCC greater than 0.5, however, the specificity was 0.5 which indicates prediction of 50% false positive regulations. In an overall, the approach did a correct estimation of NET1 structure and good approximation of the parameters.

Next we experimented with a larger network (NET2) with $N = 30$ genes to investigate the inference capability of the algorithm. The structure of the network was very sparse and it is the same architecture that was used in [21]. The parameters of NET2 were chosen arbitrarily as shown in Table 4. We performed the reconstruction experiment both in noise-free condition and 5% noise corrupted environment. The experimental conditions were once again kept the same except $I = 5$ was used to limit the maximum number of regulations for a particular gene.

It is known that with the increase of dimensionality the problem complexity increases rapidly and the GRN prediction problem is not an exception. Therefore, in predicting the correct regulations, even in ideal condition, the inference algorithm had some difficulty. In noise-free condition, the method identified more than 50% regulations correctly and identified most of the true negatives. However, the algorithm inferred many false positives and some true negatives. The summary of the prediction from noise-free condition and 5% noisy condition is

Table 5. Summary of the results of NET2 reconstruction

Measure	Noise free data	5% noisy data
Sensitivity	0.611	0.305
Specificity	0.996	0.981
Accuracy	0.981	0.954
MCC	0.7246	0.329

presented in Table 5. It is shown in Table 5 that when the gene expression data was noisy, the prediction accuracy deteriorated in terms of sensitivity and MCC. However, the specificity and accuracy was not affected much because these values were dependent on the choice of I. If we had chosen a smaller value of I, then it was possible to increase the prediction accuracy even more. However, it can be summarized that the prediction method could made an overall estimate of the network structure for a reasonably large and sparse network given sufficient gene expression data and a reasonable level of noise.

4.2   SOS DNA Repair Network Inference

We tested the proposed algorithm in the reconstruction of well-known SOS DNA repair network in *Escherichia coli*. The SOS network, consisting of 40 genes, is initiated when any damage in DNA or interference in DNA replication process is detected [24, 25]. However, the core repair system is controlled by the interplay between *RecA* and *LexA* proteins. More details about the working mechanism of the SOS DNA repair system in *E. coli* could be found in [26].

We used the gene expression data set collected in Uri Alon Lab[1]. The data set contains expression levels of 8 genes (*uvrD, lexA, umuD, recA, uvrA, uvrY, ruvA* and *polB*) of the SOS DNA repair network. Gene expression levels were measured after irradiation of the DNA with UV light. Four experiments were done for various light intensities (Exp. 1 & 2: 5 Jm^{-2}, Exp. 3 & 4:20 Jm^{-2}) in each of which 50 samples were collected at 6 minutes interval for the above 8 genes [27]. For reconstructing the network we used only the first data set and preprocessed it by ignoring the sample at first time point (which was zero) and normalizing in the range [0,1].

We identified the regulators of each gene under the same algorithmic settings except we included the decay rate as a search parameter. The search ranges were as follows: $w_{ij} \in [-10.0, 10.0], \beta_i \in [-10.0, 10.0], \tau_i \in [0.0, 10.0]$ and $\lambda_i \in [0.0, 1.0]$. The reconstruction algorithm was repeated for 10 independent trials for each gene. In each run the reconstruction process achieved a very small fitness score indicating that the estimated model could match the target time course pretty well. Fig. 1 compares the target dynamics and the estimated model generated dynamics for some selected genes of the target network. From Fig. 1 it is evident that the estimated decoupled models for the genes captured the system response adequately.

---

[1] http://www.weizmann.ac.il/mcb/UriAlon/

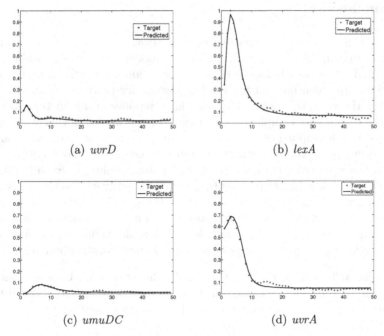

(a) *uvrD*   (b) *lexA*

(c) *umuDC*   (d) *uvrA*

Fig. 1. Target and estimated dynamics for the SOS DNA repair network (y axis represents normalized expression level and x axis represents samples at six minute intervals)

Table 6. Predicted regulatory interactions in SOS network

	uvrD	lexA	umuDC	recA	uvrA	uvrY	ruvA	polB
uvrD	+	−						
lexA	+			+				
umuDC		−	+			+		
recA	+	−						
uvrA						+		
uvrY		−		+				
ruvA		−						+
polB	+	−		+				

+ (−) represents activation (repressive) control

However, the predicted regulations and parameter values were very different from run to run in our experiments. We applied Z-score analysis to identify the robust regulators from multiple trial runs. Based on our analysis we reconstructed the network structure presented in Table 6. As shown in Table 6, the essential regulatory interactions were identified by the proposed method. Inhibitory interactions of *lexA* gene on most of the other genes were identified correctly in addition to the activation of *lexA* by *recA*. Nevertheless, the prediction also includes a number of false positives which are either unknown regulations or the side effect of noise.

5   Conclusion

Large scale gene network inference has been always impeded by the computational requirement imposed by the underlying model. Recurrent neural network (RNN) model has been found to be a good candidate for estimating the GRN from gene expression data in terms of biological flexibility and computational feasibility. However, the model contains a large number of parameter which still makes the search very complicated for large scale networks. In this work, we investigated the decoupling of the model in which the regulators of each gene are identified independently in separate search instances. We used a natural computation based search algorithm, called differential evolution, for inferring the regulators of each gene. Experimenting with two artificial GRNs and analyzing a real gene expression profile, we verified the practicability of the proposed approach. Moreover, such decoupling mechanism not only makes the identification of large networks computationally feasible but also facilitates the immediate parallelization or distributed implementation of the reconstruction algorithm.

References

1. Das, S., Caragea, D., Welch, S.M., Hsu, W.H. (eds.): Handbook of Research on Computational Methodologies in Gene Regulatory Networks, 1st edn. Medical Information Science Reference, PA (2009)
2. Zhang, Y., Xuan, J., de los Reyes, B.G., Clarke, R., Ressom, H.W.: Reverse engineering module networks by PSO-RNN hybrid modeling. BMC Genomics 10(suppl. 1), S15 (2009)
3. Gardner, T.S., di Bernardo, D., Lorenz, D., Collins, J.J.: Inferring genetic networks and identifying compound mode of action via expression profiling. Science 301(5629), 102–105 (2003)
4. Friedman, N.: Inferring cellular networks using probabilistic graphical models. Science 303(5659), 799–805 (2004)
5. D'Haeseller, P., Liang, S., Somogyi, R.: Genetic network inference: from co-expression clustering to reverse engineering. Bioinformatics 16(8), 707–726 (2000)
6. Wahde, M., Hertz, J.: Coarse-grained reverse engineering of genetic regulatory networks. Biosystems 55(1-3), 129–136 (2000)
7. Ressom, H.W., Zhang, Y., Xuan, J., Wang, Y.J., Clarke, R.: Inference of gene regulatory networks from time course gene expression data using neural networks and swarm intelligence. In: IEEE Symposium on Computational Intelligence and Bioinformatics and Computational Biology (CIBCB), pp. 435–442 (2006)
8. Wahde, M., Hertz, J.: Modeling genetic regulatory dynamics in neural development. Journal Computational Biology 8(4), 429–442 (2001)
9. Chiang, J.H., Chao, S.Y.: Modeling human cancer-related regulatory modules by GA-RNN hybrid algorithms. BMC Bioinformatics 8(91) (2007)

10. Xu, R., Wunsch II, D.C., Frank, R.L.: Inference of genetic regulatory networks with recurrent neural network models using particle swarm optimization. IEEE/ACM Transaction on Computational Biology and Bioinformatics 4(4), 681–692 (2007)
11. Noman, N., Palafox, L., Iba, H.: Inferring Genetic Networks with Recurrent Neural Network Model using Differential Evolution. In: Handbook of Bio-and Neuroinformatics - Part-C: Machine Learning Methods for Information Processing. Springer (2012)
12. Keedwell, E., Narayanan, A.: Discovering gene networks with a neural-genetic hybrid. IEEE/ACM Transaction on Computational Biology and Bioinformatics 2(3), 231–242 (2005)
13. Vohradský, J.: Neural model of the genetic network. The Journal of Biological Chemistry 276(39), 36168–36173 (2001)
14. Grimaldi, M., Visintainer, R., Jurman, G.: RegnANN: Reverse engineering gene networks using artificial neural networks. PLoS ONE 6(12), e28646 (2011)
15. Noman, N., Iba, H.: On the reconstruction of gene regulatory networks from noisy expression profiles. In: Proceedings of the World Congress on Computational Intelligence 2006, pp. 8712–8719 (July 2006)
16. Song, L., Kolar, M., Xing, E.P.: KELLER: Estimating time-varying interactions between genes. Bioinformatics 25(12), i128–i136 (2009)
17. Storn, R., Price, K.V.: Differential evolution - a simple and efficient heuristic for global optimization over continuous spaces. Journal of Global Optimization 11(4), 341–359 (1997)
18. Mondal, B.S., Sarkar, A.K., Hasan, M.M., Noman, N.: Reconstruction of gene regulatory networks using differential evolution. In: Proceedings of 13th International Conference on Computer and Information Technology (ICCIT 2010), pp. 440–445 (2010)
19. Arnone, M., Davidson, E.: The hardwiring of development: Organization and function of genomic regulatory systems. Development 124(10), 1851–1864 (1997)
20. Kikuchi, S., Tominaga, D., Arita, M., Takahashi, K., Tomita, M.: Dynamic modeling of genetic networks using genetic algorithm and S-sytem. Bioinformatics 19(5), 643–650 (2003)
21. Kimura, S., Ide, K., Kashihara, A., Kano, M., Hatakeyama, M., Masui, R., Nakagawa, N., Yokoyama, S., Kuramitsu, S., Konagaya, A.: Inference of S-system models of genetic networks using cooperative coevolutionary algorithm. Bioinformatics 21(7), 1154–1163 (2005)
22. Noman, N., Iba, H.: Reverse engineering genetic networks using evolutionary computation. Genome Informatics 16, 205–214 (2005)
23. Price, K.V., Storn, R.M., Lampinen, J.A.: Differential Evolution: A Practical Approach to Global Optimization. Springer, Heidelberg (2005)
24. Michel, B.: After 30 years of study, the bacterial SOS response still surprises us. PLoS Biology 3(7), e255 (2005)
25. Janion, C.: Some aspects of the SOS response aystem - a critical survey. Acta Biochimica Polonica 48(3), 599–610 (2001)
26. Little, J.W., Edmiston, S.H., Pacelli, L.Z., Mount, D.W.: Cleavage of the Escherichia coli lexA protein by the recA protease. Proceedings of National Academy of Science (PNAS) 77(6), 3225–3229 (1980)
27. Perrin, B.E., Ralaivola, L., Mazurie, A., Bottani, S., Mallet, J., d'Alché-Buc, F.: Gene networks inference using dynamic bayesian networks. Bioinformatics 19, 138–148 (2003)

Design and Control of Synthetic Biological Systems

Ryoji Sekine and Masayuki Yamamura

Department of Computational Intelligence and Systems Science,
Tokyo Institute of Technology, Kanagawa, Japan
green-dolphin@es.dis.titech.ac.jp,
my@dis.titech.ac.jp

Abstract. In the field of synthetic biology, genetic networks are designed by combining well-characterized genetic parts, similar to electronic circuits. Such gene networks are called synthetic genetic circuits. The design approach for synthetic genetic circuits is based on mathematical modeling and numerical simulation. The approach allows the realization of various cellular functions. However, unavoidable differences in the initial states or fluctuations of the gene expression in cells have prevented the precise prediction and control of cellular behavior. Therefore, the design of synthetic genetic circuits is not sufficient, and the dynamic control of the circuits is also required. In this report, we provide examples of synthetic circuit designs and the control of synthetic biological systems, as well as perspectives on design and control.

Keywords: Synthetic biology, Control engineering, Synthetic genetic circuits.

1   Introduction

Industrial and clinical uses of cells have recently gained attention as an important theme [1, 2]. To achieve a desired cellular behavior, the traditional (trial and error) method, in which genes are changed one by one, is inefficient. Instead, a systematic method is required, because gene networks are composed of complex combinations of genes. In the field of synthetic biology, a systematic method including numerical simulation is used to engineer gene networks that program cells to perform desired behaviors, and such gene networks are referred to as synthetic genetic circuits.

Synthetic genetic circuits are engineered by combining well-characterized genetic parts, similar to electronic circuits. The design approach of synthetic genetic circuits consists of the following three steps (Figure 1). First, a synthetic genetic circuit is designed to implement the desired cellular function in cells. Second, a mathematical model of the dynamic cellular behavior is built. Third, through numerical analysis using the model, the range of the modifiable parameter values for the desired behavior is analyzed. If the range is acceptable, then the designed synthetic genetic circuit is constructed in cells, and if not, the above three steps are repeated. If the synthetic genetic circuits constructed through the approach do not work well, we can fine-tune various parameters, such as the protein synthesis rate, which can be adjusted by mutations of the ribosome binding site (RBS) sequence, by reference to the numerical

Y. Suzuki and T. Nakagaki (Eds.): WSH 2011 and IWNC 2012, PICT 6, pp. 104–114, 2013.

analysis results. Therefore, the design approach allows the realization of novel cellular functions, such as oscillation [3-6], counts of the input numbers [7], or optimization of a metabolic pathway to produce desired materials [8]. In addition, synthetic genetic circuits that include a cell-cell communication mechanism can program cells to perform population-level behaviors, such as pattern formation [9-11] and synchronized oscillation [6].

While a synthetic circuit is useful to program cells to perform a certain behavior, a synthetic circuit cannot precisely modify the behavior, because the behavior is affected by unavoidable differences in the initial states or fluctuations of gene expression in cells. Thus, the design approach is not sufficient for applications to complex purposes, such as the induction of cell differentiation for tissue engineering, and dynamic control is also required.

In this report, we introduce examples of synthetic circuit design in Section 2 and the control of synthetic biological systems in Section 3. We will discuss the perspectives of the future integration of the design and control of synthetic biological systems in Section 4.

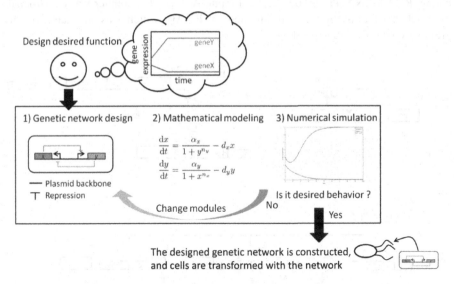

Fig. 1. Three-step approach for genetic networks to program cells to perform desired functions

2   Introduction of Synthetic Circuit Design

In this section, we introduce several synthetic circuits classified into three functional categories: switches, oscillators, and population-level behavior.

2.1   Switches

Small molecule-responsive switch-like functions are important for the rational induction of cell phenotypes in tissue engineering. Such switch-like functions have been achieved by the bistable structures of genetic networks. For example, a genetic toggle switch [12], which involved the mutual inhibition of the LacI and CIts genes (Figure 2a), was constructed in *E. coli*. Numerical analysis of the genetic toggle switch, in terms of the expression rates of the two genes, revealed that the balance of the expression rates is important for the stability. The analysis showed that the cells with the genetic toggle switch exhibited bistability *in vivo* when the expression rates of the genes were balanced. On the other hand, the cells exhibited monostability when the expression rates were imbalanced. This work was an important milestone for an integrated *in silico* and *in vivo* approach. In mammalian cells, a similar genetic network structure, named an epigenetic toggle switch [13], was constructed and consisted of the mutual inhibition of the *E. coli* erythromycin-resistance gene repressor E and the pristinamycin-induced protein PIP (Figure 2b).

A state transition in response to induction was implemented in *E. coli*. The riboregulated transcriptional cascade (RTC) counter (Figure 2c) involved the repression of RNA polymerase production, by cr binding to the RBS sequence and forming a stem-loop structure [7]. The stem-loop structure was unraveled by the non-coding RNA taRNA, driven by chemical input. A three-step counter was constructed, in which gene transcription was initiated by RNAP, which was translated by the previous count of the chemical.

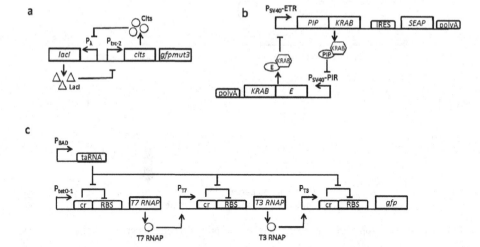

Fig. 2. Synthetic circuits programming cells to act with switch-like behavior. (a) Genetic toggle switch with a mutually inhibitory topology of the LacI and CIts. (b) Epigenetic toggle switch with a mutually inhibitory topology of the E-KRAB and PIP-KRAB proteins. (c) The RTC counter has two regulation timings. One is taRNA-regulated translation. In the absence of taRNA, the RBS on the mRNA sequence forms a stem-loop structure with cr. This structure prevents the 30S ribosomal subunit from binding to the RBS, and results in transcription inhibition. When taRNA binds to cr, cr dissociates from the RBS, and transcription occurs. The other is polymerase-regulated transcription. Sequences downstream of the T7 and T3 promoters are transcribed in the presence of the T7 and T3 polymerases, respectively.

2.2   Oscillators

Many organisms exhibit oscillatory behaviors, such as circadian rhythms and cardiac functions. The first synthetic oscillator, repressilator, consisted of three repressor proteins, LacI, TetR, and CI. LacI represses TetR production, TetR represses CI production, and CI represses LacI production [3] (Figure 3a). The three repressors are destabilized by the presence of destruction tags, because numerical simulations revealed that oscillation is likely to appear under conditions with high repressor production and a comparable degradation rate of the repressors. The period of the oscillation was longer than the doubling time of the cells, indicating that the oscillation did not occur as a consequence of cell division, but as a result of the synthetic circuit.

The dual feedback circuit includes the genes encoding the activator AraC and the repressor LacI, with expression activated by the AraC-arabinose complex and repressed by LacI. The period of the oscillation was tunable by arabinose, which drives AraC activation, and IPTG, which inhibits LacI repression [4] (Figure 3b). A theoretical report predicted that the time-delay drives oscillatory behavior [14]. Further elegant experiments then revealed that the circuit consisting of only the self-repression of LacI can oscillate, because of the time-delay of LacI folding and multimerization.

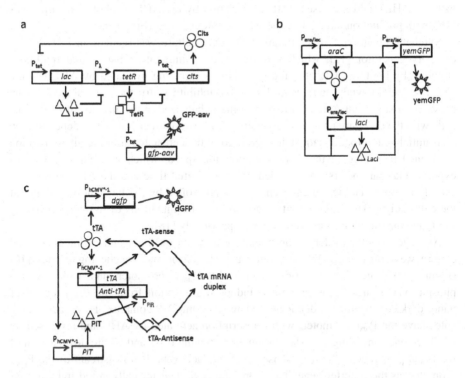

Fig. 3. Synthetic oscillators. (a) Repressilator has a three-way standoff structure composed of LacI, TetR, and CIts. (b) The dual feedback oscillator consists of positive feedback by AraC and negative feedback by LacI. (c) Synthetic mammalian clock positive feedback by tTA and transcription repression by antisense tTA mRNA. The tTA-antisense production is driven by PIT, with production positively regulated by tTA.

A synthetic oscillator for mammalian cells, a synthetic mammalian clock (Figure 3c), was achieved with the transcription activator tTA, which is translationally repressed by binding the complementary mRNA for tTA [5] (Figure 3c). The transcription of the complementary mRNA is induced by the pristinamycin-dependent transactivator, PIT, with production positively regulated by tTA. This synthetic mammalian clock, based on sense-antisense transcriptional repression and time-delay, is a representative of the molecular mechanism and the expression dynamics of a mammalian circadian clock, which presently has too many components to accurately define the molecular network.

2.3   Population Behaviors

Population behaviors by synthetic genetic circuits require cell-cell communication mediated by intercellular signaling molecules [15]. Population level behaviors in both liquid and solid cultures have been reported. Synchronized oscillation in liquid culture was implemented in *E. coli* by a quorum oscillator [6], which consisted of positive feedback by an intercellular signaling mechanism mediated by acyl-homoserine lactone (AHL), and increased AHL degradation by the AHL catabolic enzyme AiiA [16], with production induced by the AHL-LuxR complex (Figure 4a).

Synthetic phenotypic diversification in liquid culture was realized in *E. coli* by a diversity generator [17], which was designed by integrating the mutual inhibitory structure by LacI and CIts and the intercellular signaling mediated by AHL (Figure 4b). The protein synthesis rates in the mutual inhibitory structure are balanced in the presence of sufficient AHL, but they become imbalanced in the absence of AHL. The cells with the diversity generator diversified into two distinct cell states, depending on their initialization to a certain state. The ratio of the cell states after the diversification was tuned by changing the AHL-accumulation speed, which was adjusted by two experimental parameters: the cell density at the initial time and the AHL-production activity of LuxI. The initial cell density was controlled by the inoculation volume of the cell culture. The LuxI activity was modified by mutations of the gene encoding LuxI, and the mutation sites were reported previously [18].

An edge detector to detect the boundary between red light and darkness in solid culture was constructed, by combining a dark sensor and genetic logic gates [10] (Figure 4c). The dark sensor includes the P_{ompC} promoter and Cph8, which phosphorylates the OmpR protein to induce transcription from the P_{ompR} promoter under dark conditions and dephosphorylates the OmpR-P complex. The genetic logic gates have the $P_{lux-\lambda}$ promoter, with transcription activated by AHL and repressed by CI. The genes encoding the AHL production enzyme LuxI and CI are downstream of the P_{ompR} promoter in the dark sensor, and the lacZ gene is downstream of the $P_{lux-\lambda}$ promoter, as the reporter gene. Therefore, the edge detector cells in red light and far from the light did not produce LacZ, and the cells at the boundary produced LacZ.

3   Control

Unavoidable differences in the initial states or fluctuations of gene expression in cells prevent the precise manipulation of cellular behavior. Therefore, the conventional static control method called "following a recipe" is not sufficient to manipulate the higher cellular functions programmed by a synthetic genetic circuit. To achieve higher functions, such as the induction of cell differentiation for tissue engineering, a dynamic control method, in which the input amount is updated based on an observed value, is required. Thanks to the development of fluorescent microscopy for real-time measurement and micro-fluidic devices for cell incubation and chemical input, control theory is now being experimentally applied to synthetic biological systems.

Toettcher et al. achieved the control of the membrane recruitment of a protein in individual mammalian cells [19]. They constructed a red light responsive circuit, in which the fluorescent protein-fused PIF6 (BFP-PIF) binds to the membrane protein-fused PhyB (PhyB-mCherry-CAAX) under 650 nm light, and dissociates from PhyB under 750 nm light, in mammalian cells (Figure 5a). They controlled the recruitment of BFP-PIF, which could be quantified by the fluorescence intensity, by the ratio of the intensities of the 650 nm and 750 nm light inputs. The ratio was calculated by a PI controller, which is simple and powerful. The BFP fluorescence intensities of cells incubated by micro-fluidics were measured by microscopy. The cellular responses to the light inputs differed from each other, because the cells produced different amounts of BFP-PIF and PhyB-mCherry-CAAX. However, the feedback control compensated for the cell-to-cell differences.

Furthermore, feedback control of gene expression by light input has been achieved by Argeitis et al. They constructed a light-switchable gene system, in which Venus YFP is regulated by the P_{Gal1} promoter, in yeast [20]. The P_{Gal1} promoter is activated by Gal4, which consists of a binding domain (Gal4BD) and an activation domain for the P_{Gal1} promoter (Gal4AD). In the system, PhyB-fused Gal4BD (PhyB-GBD) and PIF3-fused Gal4AD (PIF3-GAD) are constitutively produced. In the presence of PCB, PhyB is converted to the Pr state under 650 nm light irradiation, and PIF3-GAD can then bind to the PhyB-GBD (Figure 5b). This binding leads to promoter activation, and results in Venus YFP production. On the other hand, when PhyB converts to the Pfr state, the bound PIF3-GAD dissociates under 750 nm light irradiation (Figure 5b). The YFP fluorescence intensity of the cells was measured by flow cytometry every 30 minutes. The intensities between the measurements were thus estimated by the use of a Kalman filter. Based on the estimated intensities, model predictive control [21] was performed to determine the input (dark, 650 nm light or 750 nm light) every 15 minutes. This feedback-control strategy achieved target intensity from various initial states, in contrast to the nonfeedback-control strategy. The feedback control of gene expression in more complex synthetic circuits seems to be challenging, because the high nonlinearity and the time delay in the relationship between an input and gene expression may prevent successful control. Indeed, Menolascina et al. tried to control CBF1-GFP production in the synthetic circuit IRMA [22], which consists of five genes, in yeast by chemical input [23], and reported open-loop preliminary experimental results.

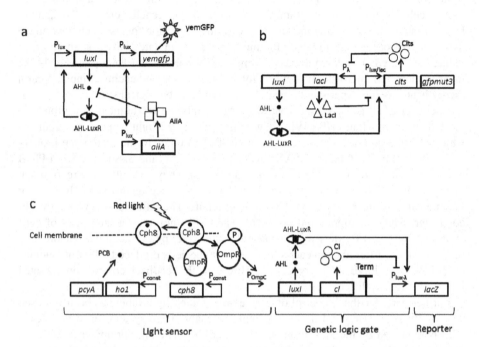

Fig. 4. Synthetic circuits for population-level behavior. (a) AHL in the population of cells with the synchronized oscillator induces the production of the AHL synthesis enzyme LuxI and the AHL catabolic enzyme AiiA. Therefore, the synchronized oscillator consists of the positive and negative feedback structures of AHL. (b) The presence of AHL in the population of cells with the diversity generator induces CIts production. In the absence of sufficient AHL, the mutual inhibitory network of LacI and CIts is bistable, whereas the network is monostable with LacI production in the absence of sufficient AHL. (c) In a dark environment, Cph8, with PCB produced by ho1 and pcyA in the edge detector, phosphorylates OmpR, which is originally produced in *E. coli*. The phosphorylated OmpR binds to the P_{OmpC} promoter and activates transcription from the promoter located upstream of the luxI and cI genes. The synthesis of the reporter protein LacZ is activated by the AHL-LuxR complex, and is repressed by CI. The edge detector cells in the dark area do not produce LacZ, because of repression by CI. The cells in the dark area produce AHL, which diffuses in the plate. Therefore, if the cells in the red light illuminated area are located near a dark area, they produce LacZ, and if not, they do not produce LacZ.

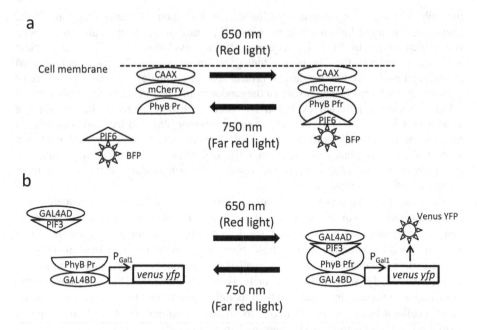

Fig. 5. Synthetic circuits that utilized dynamic control. (a) PhybB fused with CAAX-mCherry is localized to the cell membrane by the C-terminal CAAX motif of Kras. The conformation of PhyB changes from the Pr state to the Pfr state by 650-nm light irradiation, and the change can be reversed by 750 nm light irradiation. PIF6 can bind to the Pfr-state PhyB, resulting in PIF6-BFP recruitment to the cell membrane. (b) PhyB fused with GAL4BD binds to the P_{Gal1} promoter, regulating Venus YFP production. Under 650 nm light, the conformation of PhyB changes from the Pr state to the Pfr state, and can be reversed by 750 nm light irradiation. GAL4AD-PIF3 can bind to the Pfr state PhyB, and results in Venus YFP production.

4   Conclusion and Perspectives

Various synthetic genetic circuits have been developed for use in investigations of the design principles of genetic networks governing diverse biological phenomena [24] or for manipulations of biological systems, such as metabolites [1]. Since synthetic genetic circuits for higher functions require more genetic parts, the strong limitations on the number of genetic parts have hampered the development of new synthetic genetic circuits. Therefore, greater numbers of species with available genetic parts are required for larger and more complex synthetic biological systems. To increase the components of a synthetic circuit, not only repressor or activator proteins, but also non-coding RNAs and riboswitches should be developed. Indeed, non-coding RNAs [5, 7] and a riboswitch [25] have recently been used for synthetic circuits. Especially, the riboswitch will be a powerful tool for small molecule-responsive gene regulation. With an enlarged library of genetic parts, more complex functions in higher organisms will be achieved. After combining the genetic parts, tuning their parameter values, such as protein activities or protein synthesis rates, will be very important to realize a desired function. For example, the cells with the diversity generator introduced in section 2-3 cannot diversify if the enzymatic activity of LuxI is too high or low [17]. Since the parameter value of a genetic part is determined by its DNA sequence, directed evolution [26, 27] is an efficient approach to enlarge the library for tuning parameter values. In traditional directed evolution, a gene encoding the desired protein activity or the desired transcription rate of a promoter is selected from a library of the randomly mutated gene, by the application of selective pressure. Directed evolution has been applied to increase the activities of various proteins, such as enzyme [18] or signal receptor [28], and has also been utilized to increase the protein synthesis rate [29]. Further advances in directed evolution strategies, such as library construction [30], will allow us to more easily tune the parameter values of genetic parts and result in the successful construction of more complex synthetic circuits.

The computational analysis of synthetic circuit design, based on a mathematical model, provides a guide for tuning the parameters when a constructed synthetic genetic circuit does not work. Indeed, the design of the synthetic genetic toggle switch was evaluated by using parameter phase diagrams from the mathematical model [12]. Numerical simulation based on a mathematical model is frequently used for predictions of dynamic behavior, such as oscillation [3]. The selection of the information to include in a mathematical model is determined by the type of predicted desired cellular behavior. For example, to predict the formation of synthetic patterns, spatial information is included in their mathematical models [9, 10].

In addition to the design of synthetic circuits, the control of cells with the synthetic circuits is important for reliable cellular behavior. However, very few studies have applied control theory to synthetic biological systems, because control engineering has not been used for the design of genetic circuits. To apply control theory, controllability, measurability and stability must be considered in genetic circuit design. To achieve the controllability of complex genetic circuits, genetic parts such as riboswitches should be developed. Technologies such as the electroactive microwell array [31] are also required to assess more states of cells, for measurability. Furthermore, for systematic control, a linear approximation of a mathematical model of cells with a synthetic circuit, such as piecewise affine approximation [32], would be required, because a non-linear system is difficult to control systematically. The integration of synthetic biology and control engineering will generate technological breakthroughs in the fields of chemical plants and regeneration therapy.

References

1. Connor, M.R., Atsumi, S.: Synthetic biology guides biofuel production. J. Biomed. Biotechnol. 2010, 541698 (2010)
2. Ruder, W.C., Lu, T., Collins, J.J.: Synthetic biology moving into the clinic. Science 333, 1248–1252 (2011)

3. Elowitz, M.B., Leibler, S.: A synthetic oscillatory network of transcriptional regulators. Nature 403, 335–338 (2000)
4. Stricker, J., Cookson, S., Bennett, M.R., Mather, W.H., Tsimring, L.S., Hasty, J.: A fast, robust and tunable synthetic gene oscillator. Nature 456, 516–519 (2008)
5. Tigges, M., Marquez-Lago, T.T., Stelling, J., Fussenegger, M.: A tunable synthetic mammalian oscillator. Nature 457, 309–312 (2009)
6. Danino, T., Mondragón-Palomino, O., Tsimring, L., Hasty, J.: A synchronized quorum of genetic clocks. Nature 463, 326–330 (2010)
7. Friedland, A.E., Lu, T.K., Wang, X., Shi, D., Church, G., Collins, J.J.: Synthetic gene networks that count. Science 324, 1199–1202 (2009)
8. Atsumi, S., Hanai, T., Liao, J.C.: Non-fermentative pathways for synthesis of branched-chain higher alcohols as biofuels. Nature 451, 86–89 (2008)
9. Basu, S., Gerchman, Y., Collins, C.H., Arnold, F.H., Weiss, R.: A synthetic multicellular system for programmed pattern formation. Nature 434, 1130–1134 (2005)
10. Tabor, J.J., Salis, H.M., Simpson, Z.B., Chevalier, A.A., Levskaya, A., Marcotte, E.M., Voigt, C.A., Ellington, A.D.: A Synthetic Genetic Edge Detection Program. Cell 137, 1272–1281 (2009)
11. Liu, C., Fu, X., Liu, L., Ren, X., Chau, C.K., Li, S., Xiang, L., Zeng, H., Chen, G., Tang, L.H., Lenz, P., Cui, X., Huang, W., Hwa, T., Huang, J.D.: Sequential establishment of stripe patterns in an expanding cell population. Science 334, 238–241 (2011)
12. Gardner, T.S., Cantor, C.R., Collins, J.J.: Construction of a genetic toggle switch in Escherichia coli. Nature 403, 339–342 (2000)
13. Kramer, B.P., Viretta, A.U., Baba, M.D.-E., Aubel, D., Weber, W., Fussenegger, M.: An engineered epigenetic transgene switch in mammalian cells. Nat. Biotechnol. 22, 867–870 (2004)
14. Bratsun, D., Volfson, D., Tsimring, L.S., Hasty, J.: Delay-induced stochastic oscillations in gene regulation. Proc. Natl. Acad. Sci. U.S.A. 102, 14593–14598 (2005)
15. Brenner, K., You, L., Arnold, F.: Engineering microbial consortia: a new frontier in synthetic biology. Trends Biotechnol. 26, 483–489 (2008)
16. Dong, Y.H.: AiiA, an enzyme that inactivates the acylhomoserine lactone quorum-sensing signal and attenuates the virulence of Erwinia carotovora. Proc. Natl. Acad. Sci. U.S.A. 97, 3526–3531 (2000)
17. Sekine, R., Yamamura, M., Ayukawa, S., Ishimatsu, K., Akama, S., Takinoue, M., Hagiya, M., Kiga, D.: Tunable synthetic phenotypic diversification on Waddington's landscape through autonomous signaling. Proc. Natl. Acad. Sci. U.S.A. 108, 17969–17973 (2011)
18. Kambam, P.K.R., Sayut, D.J., Niu, Y., Eriksen, D.T., Sun, L.: Directed evolution of LuxI for enhanced OHHL production. Biotechnol. Bioeng. 101, 263–272 (2008)
19. Toettcher, J.E., Gong, D., Lim, W.A., Weiner, O.D.: Light-based feedback for controlling intracellular signaling dynamics. Nat. Methods 8, 837–839 (2011)
20. Milias-Argeitis, A., Summers, S., Stewart-Ornstein, J., Zuleta, I., Pincus, D., El-Samad, H., Khammash, M., Lygeros, J.: In silico feedback for in vivo regulation of a gene expression circuit. Nat. Biotechnol. 29, 1114–1116 (2011)
21. Morari, M., Lee, J.H.: Model predictive control: past, present and future. Comput. Chem. Eng. 23, 667–682 (1999)
22. Cantone, I., Marucci, L., Iorio, F., Ricci, M.A., Belcastro, V., Bansal, M., Santini, S., di Bernardo, M., di Bernardo, D., Cosma, M.P.: A yeast synthetic network for in vivo assessment of reverse-engineering and modeling approaches. Cell 137, 172–181 (2009)

23. Menolascina, F., di Bernardo, M., di Bernardo, D.: Analysis, design and implementation of a novel scheme for in-vivo control of synthetic gene regulatory networks. Automatica 47, 1265–1270 (2011)
24. Mukherji, S., van Oudenaarden, A.: Synthetic biology: understanding biological design from synthetic circuits. Nat. Rev. Gen. (2009)
25. Isaacs, F.J., Dwyer, D.J., Collins, J.J.: RNA synthetic biology. Nat. Biotechnol. 24, 545–554 (2006)
26. Yokobayashi, Y., Weiss, R., Arnold, F.H.: Directed evolution of a genetic circuit. Proc. Natl. Acad. Sci. U.S.A. 99, 16587–16591 (2002)
27. Haseltine, E.L., Arnold, F.H.: Synthetic gene circuits: design with directed evolution. Annu. Rev. Biophys. Biomol. Struct. 36, 1–19 (2007)
28. Collins, C.H., Leadbetter, J.R., Arnold, F.H.: Dual selection enhances the signaling specificity of a variant of the quorum-sensing transcriptional activator LuxR. Nat. Biotechnol. 24, 708–712 (2006)
29. Alper, H., Fischer, C., Nevoigt, E., Stephanopoulos, G.: Tuning genetic control through promoter engineering. Proc. Natl. Acad. Sci. U.S.A. 102, 12678–12683 (2005)
30. Wang, Q., Wu, H., Wang, A., Du, P., Pei, X., Li, H., Yin, X., Huang, L., Xiong, X.: Prospecting metagenomic enzyme subfamily genes for DNA family shuffling by a novel PCR-based approach. J. Biol. Chem. 285, 41509–41516 (2010)
31. Kim, S.H., Yamamoto, T., Fourmy, D., Fujii, T.: An electroactive microwell array for trapping and lysing single-bacterial cells. Biomicrofluidics 5, 24114 (2011)
32. Keshavarz, M., Barkhordari Yazdi, M., Jahed-Motlagh, M.R.: Piecewise affine modeling and control of a boiler–turbine unit. Applied Thermal Engineering 30, 781–791 (2010)

Part II
Satellite Symposium
on Computational Aesthetics

Preface: Natural Computing and Computational Aesthetics

Fuminori Akiba

Graduate School of Information Science, Nagoya University, Furocho, Chikusa,
Nagoya 464-8601, Japan
akibaf@is.nagoya-u.ac.jp

Computational aesthetics already has had a long history. As early as 1928, G.D. Birkhoff introduced the concept of the aesthetic measure (M) and defined it as the ratio between order (O) and complexity (C): M = O/C. In Japan, in September 1964, art philosopher H. Kawano published the first computer-generated works in *IBM Review* (see the Website [http://on1.zkm.de/zkm/stories/storyReader$7663] of the exhibition: "Hiroshi Kawano –The Philosopher at the Computer," 2012, ZKM, Karlsruhe, Germany).

Now there exist multiple societies for computational aesthetics such as Computational Aesthetics in Graphics, Visualization and Imaging (CAe), and the International Society for Mathematical and Computational Aesthetics (IS-MCA). Especially the latter covers a wide scope: computer-aided design and manufacturing, robot motion design, analysis of artistic masterpieces, scientific theory building and reasoning, and software design.

However, these societies concentrate their attention almost exclusively on aesthetically designed objects or on designing objects aesthetically, even though their compass ranges from automobile to scientific theories. To us this seems somehow one-sided.

This section contains three chapters. Each offers a point of view different from that of already existing computational aesthetics. Akiba proposes computational aesthetics of "nature." In order to do so, he retrieves the wider scope of Kant's aesthetics in *Kritik der Urteilskraft* from the narrow interpretation made by existing computational aesthetics; he points out that in the idea of "harnessing" in natural computing we can find a successor of Kant's aesthetics and a possibility of computational aesthetics of nature.

Goan et al., on the basis of G. Bateson's learning theory (stepping up from logical types in a dead-end situation) and J.J. Gibson's concept of ambient space, and through the elaborate workshop at the art museum, show that "there could exist a way of perceiving the ground–ground switch, the perception of surfaces' layouts, by stepping up from the logical type of figure–ground reversal perception—the figure–figure switch." At the same time it also shows the critical responses to Akiba's idea of computational aesthetics of nature and to the idea of "indirect control" in natural computing.

Y. Suzuki and T. Nakagaki (Eds.): WSH 2011 and IWNC 2012, PICT 6, pp. 117–118, 2013.

Watanabe introduces us to unique interfaces that he and his colleagues developed, such as "Saccade-Based Displays," "Save YourSelf!!! [Galvanic vestibular stimulation]," and the workshop called "Heartbeat Picnic." Experience mediated by such interface technologies "induces appreciation about self, and makes us aware of new rules as to how people relate to their environments." He also relates self-awareness experience to the idea of Kant's aesthetic judgment, formal purposiveness, and subjective universality, opening up another computational aesthetic of human nature for us.

Of course, these chapters contain much more than what has been mentioned here in passing. We hope readers will find further possibilities in this work to develop the future relationship between natural computing and computational aesthetics.

The Significance of Natural Computing for Considering Computational Aesthetics of Nature

Fuminori Akiba

Graduate School of Information Science, Nagoya University, Japan
akibaf@is.nagoya-u.ac.jp

Abstract. In this paper, I aim to propose computational aesthetics of nature and to explain how the idea of "harness" in natural computing is centrally important to this end. First, I reconfirm the original scope of Kant's aesthetics that is at the core of computational aesthetics. In the discussion of Kant's aesthetics, I contend that the reason Kant introduced the concepts of beauty in nature and aesthetic judgment into his philosophy is because he recognized them as key drivers behind the development and cultivation of our understanding of nature. I point out that the field of computational aesthetics presently understands Kant's aesthetics only insufficiently. There thus exists a need to propose another computational aesthetics of nature and to define its central aim: to find the beautiful in nature, leading us both to a better understanding of nature, and to a greater awareness of how we should live in nature. With this aim in mind, "harness" emerges as a useful tool for computational aesthetics of nature—one that merits consideration.

Keywords: aesthetics, computation, nature, Kant, harness, indirect interaction, tactile score.

1   Introduction

Since G.D. Birkhoff (1928), computational aesthetics (CA) has had a long history (Scha & Bod 1993, Hoenig 2005). Correspondingly, there now exist multiple societies devoted to the research and promotion of CA, Computational Aesthetics in Graphics, Visualization and Imaging (CAe), and the International Society for Mathematical and Computational Aesthetics (IS-MCA)—the latter of which engages with a seemingly wide variety of topics and subfields (see the website of IS-MCA*).

In spite of a variety of topics, in their background we can easily find the strong echoes of a philosopher—Immanuel Kant (1724-1804). For example, IS-MCA website says "all design attempts to satisfy two constraints: functionality and aesthetics. Even a discipline as functionally oriented as structural engineering, in fact, involves aesthetic control over systems of non-linear equations". It easily reminds me one of famous ideas of Kant's aesthetics, "formal purposiveness" which I explain in 2.3 below.

Other researchers, explicitly or implicitly, also invoke Kant's aesthetics. "The best analysis of the esthetics [sic] is still Immanuel Kant's. He viewed experience of beauty

Y. Suzuki and T. Nakagaki (Eds.): WSH 2011 and IWNC 2012, PICT 6, pp. 119–129, 2013.

as the consciousness of a psychological process: the pleasing awareness of harmony in the free play of our cognitive faculties" (Scha and Bod 1993). "Kant had also described aesthetics as a reinforcing supplement to logic ideas or concepts hinting that objects are higher value to us if they are beautiful" (Hoenig 2005). Bertelesen 2002 says "[T]he problem [of aesthetic computing] is the paradoxical one of meeting needs that don't yet exist, supporting the development of practice that we cannot yet imagine." The phrase "meeting needs that don't yet exist" or "supporting the development of practice that we cannot yet imagine" suggest correspondence to the key concept of Kant's aesthetics, "formal purposiveness" again.

Yet a number of CA scholars appear to unduly limit the ways in which Kant's aesthetics may be applied and understood. Hoenig (2005), for example, identifies essential questions: "Can we construct tools that assist with creating beauty as easily as they do now with purely functional development? Can we make machines aware of aesthetics in a similar fashion as humans are?" The narrow, "industrial" scope of the abovementioned questions is unsettling. IS-MCA is also concerned only with "design object", even though the idea of design object covers a wide variety of subjects ranging from "the machine-sculpted surface of a car body" to "the Feynman propagator in quantum electrodynamics."

The need for CA's rapid practical applicability to industrial design understandably promotes such a limited approach. However, is it the primary task of CA to help design beautiful artificial objects, which have aesthetic qualities beyond mere functionality, with computational methods? Is it the main task of aesthetics to develop supporting tools for designing such beautiful objects?

The answer to each of the above is probably "no," even though such tasks are significant. The scope of the field portended by Kant's aesthetics—a philosophical system inherently embedded in the understanding of computational aesthetics—is much wider. As we will see in the following sections, the reason why Kant introduced the ideas of beauty in nature and aesthetic judgment into his philosophy is because he recognized these two concepts as key drivers behind the development and cultivation of our understanding of *nature*. Of course, this is not to say that CA's present (limited) understanding of Kant's aesthetics is wrong; for, in fact, it does call on a part of Kant's aesthetics. However when we consider the entire scope of Kant's aesthetic theory, we realize that CA's present understanding of Kant's aesthetics is insufficient, and that we should propose another computational aesthetics of nature.

2   The Scope of Aesthetics Suggested by Kant's *Critique of the Power of Judgment*

Here I mention only basic points indispensable for recovering the wide-scope approach that I can derive from Kant's philosophical aesthetics. The basic points are as follows:

1) Kant introduced the idea of "purposiveness" as a principle of judgment in order to advance our understanding of nature (2.1);

2) On the basis of a principle of purposiveness, "reflective judgment" advances our understanding of nature (2.2);
3) Aesthetic judgment, as a kind of reflective judgment, cultivates our interest in nature. Beauty and aesthetic judgment facilitate our understanding of nature (2.3);
4) In addition, beauty and aesthetic judgment also cultivate our moral feelings (2.4);

At the end of this section, I confirm the roles of the beautiful in nature and aesthetic judgment, and point out that CA's understanding of Kant's aesthetics is insufficient. Finally, I define the essential task of aesthetics different from presently accepted (2.5).

2.1   The Principle of the Lawful Unity of Nature

For Kant, the most important thing is to understand nature as mechanism. He assigned the role of advancing our understanding of nature to the faculty of judgment. Through the power of judgment (more precisely, what Kant deems our "determinant judgment"), we subsume a specific natural object before us under the laws of nature as mechanism. However, when the diversity or multiplicity of nature goes too far, it becomes harder for our judgment to determine under which specific laws the objects of nature before us should be subsumed. Judgment loses its way. Even when the differences between natural objects are perceived to be great, we must find higher laws that interconnect them, and thus continue to develop our understanding of nature. But how can we do so? This is at least one of the reasons Kant wrote his treatise, the *Critique of the Power of Judgment.* He tries to rescue judgment from such difficulty and gives another potent principle to judgment reflecting upon nature. It is a principle that holds the "lawful unity" of nature (Kant 70). We can assume that the great, confounding multiplicity nevertheless has unity under a few principles.

2.2   A Principle of Purposiveness and Two Kinds of Judgment

2.2.1   A Principle of Purposiveness for Cognitive Faculties

To this "lawful unity", Kant gives another name. He calls it "a principle of purposiveness for our faculty of cognition" (70-71), because if we are allowed to assume that nature contains a lawful unity, it is quite purposive for our cognitive faculties whose purpose is to improve our understanding of nature.

Under the guidance of this principle, the power of judgment tries to find "possible (still to be discovered)" (70-71) laws of nature. However, in this case, as he says, the principle is requested by judgment itself for its own guidance, for its own cognitive faculty, not for the sake of the object or objects before it. Therefore, Kant calls it the "subjective" principle of judgment.

2.2.2   Two Principles and Two Kinds of Judgment Requested to Advance Our Understanding of Nature

Here we recognize that there are two kinds of judgment. One is the kind of judgment that subsumes specific objects of nature under pre-established laws—what Kant calls a "determinant judgment," governed by a "constructive" principle. The other type of judgment is one which reflects upon nature without laws [and concepts], and tries to find possible (still discoverable) laws of nature. Kant dubs this "reflective judgment," operating under the "regulative" principle. On the basis of this classification, Kant defines the principle of purposiveness for the cognitive faculty as the regulative and subjective principle for "reflective" judgment. We utilize these two complementarily to advance our understanding of nature.

2.3   The Status and the Role of Aesthetic Judgment

As I have just remarked, one of the reasons Kant wrote his *Critique of the Power of Judgment* is to advance our understanding of nature. And it is in this context that he introduces aesthetic judgment. In order to realize the significance and the role of aesthetic judgment in this context, the following three points are important:

1) Aesthetic judgment is the purest form of reflecting judgment (see 2.3.1 below);

2) Aesthetic judgment has the same subjective condition as all judgments. However, in judging beauty in nature, our cognitive faculties stay in a state of free play without achieving practical ends and keep animating one other (2.3.2);

3) This animation resulting from judging beauty in nature facilitates our attention to nature. (2.3.3)

2.3.1   Aesthetic Judgment Is the Purest Form of Reflective Judgment: A Principle of "Formal" Purposiveness for Cognitive Faculties

As a kind of reflective judgment, it does not subsume natural objects under already established laws. However, in contrast to reflective judgments in general, it does *not* try to find a (still to be discovered) rule *outside* the judging subject. It relates natural objects only to the judging subject's feelings of pleasure or displeasure. The subjective feeling of pleasure or displeasure conditions the rule: If an object results in pleasure for someone who judges it, and satisfies him or her without any specific purpose, then the object is beautiful; if it does not, it is not beautiful.

As we have seen, reflective judgments in general rely on the principle of purposiveness, with the aim of improving our understanding of nature. In this sense, it relates the principle still to objects in nature. In contrast, aesthetic judgment relates it only to the feeling of pleasure. It completely lacks substantial aims. Kant thus describes this character of aesthetic judgment as "formal," the opposite of substantial, and thus deems the principle of aesthetic judgment as one which has as its basis the "formal purposiveness of nature." Aesthetic judgment contains a principle for reflecting upon nature in its purest form:

In a critique of the power of judgment the part that contains the aesthetic power of judgment is essential, since this alone contains a principle that the power of judgment lays at the basis of its reflection on nature entirely *a priori*, namely that of a formal purposiveness of nature in accordance with its particular (empirical) laws for our faculty of cognition, without which the understanding could not find itself in it... (79)

2.3.2 Aesthetic Judgment Has the Same Subjective Condition as All Judgments. However, in Judging the Beauty in Nature Our Cognitive Faculties Stay in a State of Animated Free Play without Achieving Practical Ends

Kant says that, with regard to the representation of an object, the agreement of imagination and understanding is required as the subjective condition for all judgments.

The subjective condition of all judgments is the faculty for judging itself, or the power of judgment. This, employed with regard to a representation by means of which an object is given, requires the agreement of two powers of representation: namely, the imagination (for the intuition and the composition of the manifold of intuition), and the understanding [in German, *Verstand*] (for the concept as representation of the unity of this composition). (167)

Concerning aesthetic judgment the subjective condition is the same. However, there is a significant difference. Normally, understanding dominates imagination and unifies the manifold of intuitions given by imagination into the concept given by understanding. In contrast, when judging the beautiful in nature, imagination does not obey understanding. They mutually animate one another, because aesthetic judgment lacks any laws, under which the object should be subsumed, and any concepts, under which the manifold of intuition should be united. Kant writes:

> [S]ince the freedom of the imagination consists precisely in the fact that it schematizes without a concept, the judgment of taste [i.e., aesthetic judgment] must rest on a mere sensation of the *reciprocally animating* imagination in its freedom and the understanding with its lawfulness, thus on a feeling that allows the object to be judged in accordance with the purposiveness of the representation (by means of which an object is given) for the promotion of the faculty of cognition in its free play... (167), italic by Akiba

In short, aesthetic judgment is judgment which arises from the minimum subjective condition of all judgments, and which animates itself through the act of judging the subjective purposiveness of the representation before it.

2.3.3 The Animation of Our Cognitive Faculties in Judging the Beauty in Nature Facilitates Our Attention to Nature

Consequently, the more variation the representation contains, the more lasting this animation. And the thing that contains the widest variety in the world is, of course, nature. Therefore, aesthetic judgment is inevitably fascinated by nature: "...[N]ature, which is there extravagant in its varieties to the point of opulence, subject to no coercion from artificial rules, could provide his taste with lasting nourishment" (126). In judging "a free and indeterminate purposiveness" of the natural objects, "the mental powers" (i.e., understanding and imagination) are entertained "with that which we call beautiful" (123).

In this way the beautiful in nature—and aesthetic judgment, which is fascinated by it—cultivate our interest in nature and make us attentive to it. It is also important that Kant thinks the beautiful in nature "cultivates a certain liberality in the manner of thinking" (as below) because it brings us a possibility to see nature in a different way:

"The beautiful prepares us to love something, even nature, without interest…the beautiful in nature likewise presupposes and cultivates a certain liberality in the manner of thinking, i.e., independence of the satisfaction from mere sensory enjoyment, nevertheless by means of it freedom is represented more as in play than as subject to a lawful business…"(151)

To briefly summarize the discussion above: in order to advance our understanding of nature, Kant introduces two different but complementary principles. One is a so-called constructive principle for determinant judgment, and the other is a regulative principle for reflective judgment, that is, a principle of purposiveness. On the basis of the latter principle, the power of aesthetic judgment judges the beautiful in nature and cultivates our interest in the diversity of nature. Thus, beauty and aesthetic judgment play significant roles as motivators, or rather cultivators, of our understanding of nature (Table 1).

Table 1. Classification of judgment in Kant's *Critique of the Power of Judgment*

judgment		its principle	is related to	its role	its view of nature
determinant		mechanism	object(s) in nature	constructive	mechanical
reflective	reflective in general	purposiveness	the subject, but indirectly to object(s) in nature	regulative	teleological
	aesthetic	subjective/ formal purposiveness	subjective feelings	facilitative	----

2.3.4   Beauty and the Moral Value

One more thing we must not forget is that Kant assumes a strong relationship between the beauty of nature and moral value. Kant makes the assertion "that to take an immediate interest in the beauty of nature […] is always a mark of a good soul, and that if this interest is habitual, it at least indicates a disposition of the mind that is favorable to the moral feeling, if it is gladly combined with the viewing of nature" (178).

The principle of purposiveness is also the principle of teleological judgment. By this principle, we assume that nature organizes itself as if it prepares for the realization of a final purpose [*telos*], that is, the realization of moral goodness. This assumption is supported by the existence of organisms that produce themselves as if they had a purpose in and of themselves, and by the existence of human beings who are obliged to realize a final purpose in the world. Beauty in nature and aesthetic judgment, through the principle of purposiveness that they share with teleological (moral) judgment, make us attentive to such moral values.

2.4   CA Understands Kant's Aesthetics Partially, But Insufficiently

CA's understanding of Kant's aesthetics is not wrong because it does, in fact, depend upon a part of Kant's aesthetics. However, given the entire scope of Kant's aesthetics as discussed above, current CA's understanding of Kant's aesthetics is insufficient because it is not inclusive of the fact that Kant recognizes beauty in nature and the power of aesthetic judgment as cultivators, which further the development of our understanding of nature and its moral values. Therefore, we should propose another computational aesthetics of nature.

2.5   The Essential Task of Computational Aesthetics of Nature

From what we have discussed above, we may say that:

1) Beauty must facilitate our understanding of nature;
2) At the same time, it must facilitate our moral feelings, in other words, make us aware of how we should live in nature.

It does not matter at all whether such beauty takes the form of artwork or designed objects.

In this light, we also must return to the definition of aesthetics of nature. *Aesthetics of nature is the science of beauty in nature.* And the essential task of aesthetics of nature is *to find beauty in nature which advances our understanding of nature and which makes us aware of how we should live in nature.* In order to achieve this aim, aesthetics of nature utilizes *two different principles.* One is a scientific (mechanical) principle. The other is a complement to it. If computational aesthetics is to be any kind of aesthetics at all, it must succeed in using these principles to find beauty in nature.

3   Natural Computing and Its Significance for Computational Aesthetics of Nature

From this standpoint, natural computing (NC)—and its concept of "harness"—shows the direction in which CA must turn, because NC does not only aim to make artificial things, but also to advance our understanding of nature through computing.

3.1   Two Ideas of Computation and the Idea of "Harness"

3.1.1   Two Ideas of Computation: Constructive and Oracle
NC emerges as highly important because it is an approach that uses two different ideas of computation in order to advance natural science: "constructive computation" and "oracle computation." The constructive type refers to when we can explicitly show every step of an algorithm. Oracle computation, by direct contrast, describes when we cannot explicitly show every step of an algorithm (Suzuki 2009), such as is the case with *Physarum* computation.

Table 2. Two ideas of computation: constructive and oracle

computation	its algorithm	its representative
constructive	explicit	TM (Turing machine)
oracle	implicit	DNA, Physarum

In the following, we introduce the idea of "harness" in NC and its relationship to these two kinds of computation, and then we point out its significance for the project of computational aesthetics of nature.

3.1.2   The Idea of "Harness"

In its original sense, the word "harness" means "a set of strips of leather and metal pieces that is put around horse's head and body so that the horse can be controlled and fastened to a carriage, etc." (*Oxford Advanced Dictionary*). In its more general usage, the term has come to represent the way human beings have corralled and utilized natural forces to perform work.

In the sphere of NC, the word "harness'" is used in a slightly different way. It simply denotes alternate means of control. We shall explain it by way of comparison with the traditional method of control. With the traditional control method, we strictly limit the boundaries of a system and directly manage its given inputs and outputs. In order to do so, we must know every step of the system, one by one. Let us call this the "direct" type of control. Consider, however, that a shepherd walking with a picture of a wolf in his hands can move a flock of sheep to the pen. There is no need to know every detailed mechanism of the process, no need to manage each sheep directly and individually, and no need to set strict boundaries. The only one thing the shepherd must know is the fact that he can move a natural system (in this case, a flock of sheep) with an artificial thing (in this case, a picture of a wolf). This is what may be dubbed the "indirect" type of control, which corresponds to the idea of the "harness."

Take a tritrophic [plants-harmful insects (herbivores)-natural predators (carnivores)] ecosystem as another example. From the perspective of direct control, the best way to avoid harmful insect damage in this system is to launch direct attacks against the perpetrating insects with agrochemicals. However, this unnatural extinction completely destroys the balance of the ecosystem. Therefore, if nonnative insects that locally lack natural predators enter the picture, the entire ecosystem dies. On the contrary, nature does not do such foolish things. It utilizes indirect control via what are called Herbivore Induced Plant Volatiles (HIPVs) to regulate interactions between plants, insects, and predators. When insects eat plants, plants produce HIPVs that act as signals to natural predators (such as birds) that their prey is close at hand. Because of this indirect system of control, the ecosystem can avoid unnecessary extinctions and sustain stable diversity. It is completely different from direct control methods, which can effect local and total extinctions, and from direct food chain systems, which can often result in unstable population changes (see Suzuki and Sakai 2012 in this volume).

If we can imitate the indirect interactions inherent to the natural system—in other words, if we can harness the natural system—we can contribute to sustaining its diversity. However, in order to do so, we must know about the existence of such indirect interactions in nature, know what factors drive the interactions, and know how to construct information pathways such as the one modeled above in the HIPVs example.

3.1.3 The Idea of Harness Encourages Us to Find Possible (Still Discoverable) Indirect Interactions in Natural Systems: Harness and Two Kinds of Computation

Consequently, the idea of harness requires us to study natural systems because we need to find possible (still discoverable) indirect interactions in natural systems. It promotes our understanding of nature. However, we do not have to know every detailed algorithm of a given natural system. Nature is so complex that we may never know every step to its processes. In this sense, natural systems remain examples of oracle computations. However, if we successfully find accessible points, such as are represented by HIPVs, that enable us to understand and operate indirect interactions, we can glean clues towards comprehending the overall oracle computational system. We can thereby construct artificial materials and insert them into natural systems.

Now we can define the idea of harness more precisely, from the perspective of computation. Harness provides an alternate means of control that operates oracle computations by using the products of "constructive" computations. On the basis of this idea, researchers in the field of NC study, for example, the modifications of influenza and info-chemical signals in chemical ecology (see articles in this volume).

3.1.4 The Significance of Harness for Aesthetics of Nature

Probably the reader has already noticed that there are correspondences between the idea of harness in NC and aesthetics of nature. Firstly, the concept of harness in NC advances our understanding of nature. In addition, it facilitates our moral or ethical feelings for nature. As we have just seen above, in the idea of harness we find a clear sense of moral (or ethical) duty to "sustain the diversity in natural systems." From these two points, it follows that the idea of harness in NC suggests a desirable direction for the discipline of computational aesthetics of nature.

3.2 Human Nature and the Harness Concept in NC

At the end of this paper, we briefly mention research on human nature from the standpoint of the harness concept in NC. If harness is an attempt to operate oracle computations through constructive computations on the basis of (still to be discovered) interactions, it can be likewise applicable to the mysterious systems which comprise human nature.

3.2.1 Massage and Tactile Score

The tactile sense, or sense of touch, is a sense for which the sciences still have not provided sufficient explanations. Of course, we all know from numerous reports that massages (offering stimulation via the tactile sense) often relieve various disorders, for example, skin disorders. A multitude of studies have tried to make clear what makes such relief possible. Yet unfortunately, we still do not have enough evidence regarding the matter –it corresponds to oracle computation—. And the algorithm of interaction between touching and touched still remains implicit. If we could find some necessary (still to be discovered) access points in the interactions between touching and touched, and if we could operate the interaction by way of accessing these points, then we could give assistance to people suffering from certain disorders.

In relation to this point, a significant project is underway concerning tactility. It invents the concept of a "tactile score," and through scientific analysis, gradually reveals the existence of formal structures of massages (Suzuki, Watanabe, & Suzuki 2012). In the future, it might be possible for us to find accessible points in these structures. Those people who misunderstand massages as the stuff of hedonism and think of them not as the subject of aesthetics, do not understand the scope of Kant's aesthetics – even though they may believe that they depend on Kantian formalist aesthetics.

3.2.2   Media through Which We Become Aware of Our Nature and How to Live with Them

Just as the tactile score tries to reveal what is behind our sense of touch, and tries to make us think how should we use it, certain media attempt to clarify principles of human nature and to determine how we should transform our lives accordingly. For example, *Saccade-based Display* (Ando & Watanabe) is a medium which brings us such experiences as "consist of deterministic information systems aimed for triggering perceptual experience, which can open up questions about what human beings are, and why they perceive in such ways" (Watanabe 2012, see also his article in this volume, Cf. Akiba 2003). From this point of view we can reinterpret "designed objects" in the field of existing CA, and learn many things for the development of computational aesthetics of nature.

4   Conclusion

The reason why Kant introduced beauty in nature and aesthetic judgment in his philosophy is because he recognizes them as cultivators behind the development of our understanding of *nature*. Given the entire scope of his aesthetics, current computational aesthetics is insufficient and we must propose another computational aesthetics of nature. Its task should be to find the beautiful in nature—a pursuit which leads both to a better understanding of nature, and to a greater awareness of how we should live in nature. From this perspective, we can appreciate the notion of harness in the practice of NC as a worthwhile pursuit in furtherance of this end, and a signpost that signals a desirable direction for computational aesthetics.

Acknowledgement. This work was supported by JSPS KAKANHI Grant Number 21520133 & 24520106.

References

Akiba, F.: Information and Expression. Selected Papers of the 15th International Congress of Aesthetics, Makuhari, pp. 3–6 (2003)

Bertelsen: Aesthetics as Means for Supporting Development in Use - Beyond the Designed Purposefulness. In: Bertelsen, O., Fishwick, P. (eds.) Aesthetic Computing, Dagstuhl Seminar Report No. 348, http://www.dagstuhl.de/Reports/02/02291.pdf

Hoenig, F.: Defining Computational Aesthetics. In: Neumann, L., Sibert, M., Gooch, B., Purgathofer, W. (eds.) Computational Aesthetics in Graphics, Visualization and Imaging, Eurographics Digital Library (2005), http://diglib.eg.org

IS-MCA (September 20, 2010),
http://www.rci.rutgers.edu/~mleyton/ISMA.htm

Kant, I.: Critique of the Power of Judgment (1790); trans. Guyer, P., Matthews, E.: Cambridge University Press (2001)

Scha, R., Bod, R.: Computational Esthetics. In: Informatie en Informatiebeleid 11(1), pp. 54–63 (1993) http://iaaa.nl/rs/compestE.html

Suzuki, Y.: Self-Organization is Computation. In: Handbook of Self-Organization, S.T.N. (2009) (in Japanese)

Suzuki, Y., Watanabe, J., Suzuki, R.: Tactile Score, a Knowledge Media of Tactile Sense for Creativity. In: Watanabe, T., Watada, J., Takahashi, N., Howlett, R.J., Jain, L.C. (eds.) Intelligent Interactive Multimedia: Systems & Services. SIST, vol. 14, pp. 579–587. Springer, Heidelberg (2012)

Watanabe, J.: An Interpretation of New Media Experiences Motivated by the Viewpoint of Computational Aesthetics. In: Proceedings of 6th International Workshop on Natural Computing (JSAI 2011) (2012)

Perceiving the Gap: Asynchronous Coordination of Plural Algorithms and Disconnected Logical Types in Ambient Space

Miki Goan[1], Katsuyoshi Tsujita[2], Takuma Ishikawa[3], Shinichi Takashima[4], Susumu Kihara[4], and Kenjiro Okazaki[4]

[1] Osaka City University,
3-3-138 Sugimoto Sumiyoshi-ku, Osaka 558-8585, Japan
goan@kmu-f.jp
[2] Osaka Institute of Technology,
5-16-1 Ohmiya, Asahi-ku, Osaka 535-8585, Japan
tsujita@bme.oit.ac.jp
[3] Ochabi Institute,
2-3 Kanda-surugadai, Chiyoda-ku, Tokyo 101-0062, Japan
takumaro2001@hotmail.com
[4] Kinki University, International Center for Human Sciences Yotsuya Art Studium,
Yotsuya 1-5, Shinjuku-ku, Tokyo 160-0004, Japan
{shintakshintak,susumu.kihara,okazakipark}@gmail.com

Abstract. The purpose of this study is to elaborate on the concept of "ambient space," the space which surrounds and moves with the self. For this purpose, this study focuses on the learning processes of the creation and viewing of art, related to both artists and viewers. Furthermore, we intend to deepen thought about ambient space and reconsider the dichotomy of "figure and ground." Through a workshop at an art museum, we showed that by stepping up from the logical type of figure-ground reversal perception - the figure-figure switch - the ground-ground switch, the perception of surfaces' layouts, can be perceived.

1 A Step Up in Logical Type

When we see evidence of distinguished skills, even if we cannot explain the excellence of the skill in words we can still be aware of it. Bateson's learning theory discusses the essential dilemmas and difficulties related to skill learning and the transmission of skills. As Bateson pointed out, one dilemma faced by artists is about a double effect of practice: "It makes him, on the one hand, more able to do whatever it is he is attempting; and, on the other hand, by the phenomenon of habit formation, it makes him less aware of how he does it" (Bateson, 1972, p.138). Another dilemma is that "if his attempt is to communicate about the unconscious components of his performance, then it follows that he is on a sort of moving stairway (or escalator) about whose position he is trying to communicate but whose movement is itself a function of his efforts to communicate" (op. cit., p. 138). Bateson deeply understood "that consciousness is necessarily

Y. Suzuki and T. Nakagaki (Eds.): WSH 2011 and IWNC 2012, PICT 6, pp. 130–147, 2013.
© The Author(s) 2013

selective and partial, i.e., that the content of consciousness is, at best, a small part of truth about the self" (op. cit., p. 144). However, we agree with Bateson that the unconscious, as postulated by Freud, is not essentially unknowable. The framework he proposed assumed the existence of meta-level contexts (an invisible frame) that cover the algorithms in the area of so-called unconsciousness. In other words, the framework assumes the existence of things that enable measurement and judgment only as a function of logical typing. Bateson was a key figure in pointing out the reliance on the invisible and unconscious meta-frame and the great belief in the capability to step up logical types when we are in a dead-end situation, such as a double bind or an antinomy. He clearly pointed out this belief they can be a condition for learning.

2 Ambient Space

The purpose of this study is to elaborate on the concept of "ambient space," the importance of which was emphasized not only by G. Bateson but also by ecological psychologist J. J. Gibson and the painter J. Pollock, who developed their ecological views at around the same time. Ambient space refers to the space which surrounds and moves with the self. Bateson would say that it was the invisible context which is organised by logical types which can be recombined in several ways. For Gibson, it would be an environment whose structure would be specified by "ambient optic arrays," which refers to ambient light with structure. Ambient light means that "light would come to every point; it would surround every point; it would be environing at every point" (Gibson, 1979, p. 51). Light fills in so as to illuminate the structure of the environment, scattering and running against the myriad surfaces of objects in ambient space. "This implies an arrangement of some sort, that is, a pattern, a texture, or a configuration" (op. cit., p. 51). In Gibson's ground theory (1950), "space was no longer described as a void containing detached objects, but rather as intersecting surfaces; space was structurally more complex and differently organized"(Lombardo, 1987, p. 24). That is, Gibson denied Newton's concept of empty space, which declared that "the physical spatial world consisted of "objects" (static, self-contained things) possessing size, shape and location within space (a void)" (op. cit., p. 24). In ambient space, when we think of the relation between self and the ambient light which surrounds it, for example, it is necessary not to think first of the "things" which are related (the "relata") such as "self" and "light", but rather the opposite, so "the relations are to be thought of as somehow primary, the relata as secondary." "Beyond this, it is claimed that the relations are of the sort generated by processes of information exchange" (op. cit., p.154). Additionally, Lombardo (1987) indicated that Gibson's ecological concept of the ambient optic array (which is that reflected light diverges outward into the medium, but at any given point in the medium, light is converging from surrounding (ambient) objects) is very similar to Leibnitz's "monad":

described as the embodiment of the surrounding whole within the part, an entity "mirrors" its relationships to all surrounding entities. The monad provided an alternative building block to Newton's atom, offering a relational view of reality. Instead of intrinsically independent atoms (elements), each of which is internally simple, the relational view rejects the idea of absolute parts independent of the whole" (op. cit., p. 52).

The artist Jackson Pollock expended great mental energy and strength to describe the ambient space. As a learner, Pollock analyzed the ambient space surrounding his body, invented painting techniques such as dripping and pouring, and attempted to reconstruct the ambient space on the canvas, the medium itself. However, generally speaking, describing and analyzing the ambient space is difficult. Although the idea of the ambient space that Gibson defined and Pollock demonstrated denies explanation by way of an analogy between the retinal image and the picture by using the pictorial model, regrettably even paintings by Pollock, who studied ambient space, are treated as finished products for image analysis. With all methods of fractal analyses for complex systems (Taylor et al., 2000), the idea that many hierarchical layers of arbitrary shapes are regarded as "ground" and the segments of "lines" intersecting hierarchical blocks, on the other hand, are regarded as "figures" is nothing but a tenet of image analysis. In the pictorial model, optical stimulation is regarded as having properties of momentarily fixed conditions; that is, a series of separate static pictures. However, the medium filled with light in ambient space does not form into an "image." Lombardo (1987, p. 39) explained that

Aristotle sees the medium as revealing forms rather than transmitting something material (*eidola*). The medium is a field actualized through light. It is not a void or an empty space — it possesses the power of "transparency". The perceiver sees forms through the medium; the perceiver does not see the medium... as Gibson would say, the perceiver does not see "light" —the perceiver sees by means of light.

3   Focus on the Creation Process (Learning Process) of the Arts: Reconsideration of the Dichotomy of "Figure and Ground"

Art is commonly concerned with learning, i.e., with bridging the gap between the more or less unconscious premises acquired by learning and the more episodic content of consciousness and immediate action (Bateson, 1972, p.308). This study focuses on the learning processes of creation and viewing, related to both artists and viewers, in order to elaborate the idea of ambient space. As an example, imagine that someone is asked to remember the assignment of white and black go stones on a go board after the final move? It is very difficult to guess correctly whether a stone is white or black by tracing the position on the go board from one corner in turns. It also has a high usage of memory resources.

There is an easier way which is well known to the people learning to play go. This is simply to recall each move from the beginning in turns, and to assign the go stones to positions based on the moves. This study focuses on such a learning process. The learning process which is the significant focus of this study involves re-experiencing our own and others' learning processes. The learner need not repeat any previous misunderstandings or settle for the top-down lectures of specialists such as art historians or art critics, which are external to personal memories and experiences.

Another aim of this study is to deepen thought about ambient space, reconsidering the dichotomy of "figure and ground," which Gestalt psychology proposed. Considering a space as "the ambient space" that is filled with medium of the ambient optic array, the binary division into either figure or ground will be revealed to be insufficient to express such a space (Gibson, 1979). It is usually considered that "ground" is the expression of medium and is behind the "figure," the expression of the object. However, according to the definition of ambient space, the medium surrounds the object not only behind but also to the fore, that is, the whole. When seen from the perspective of ambient space, the idea of "figure-ground reversal" (Rubin, 1915) becomes insignificant because the idea is nothing but to return to a superficial viewing of difference, keeping the propriety claim of the dichotomy. But we don't intend to completely deny the dichotomy of figure and ground in the expression of space, insisting that it is too rough and of no use. We would like to verify the problem of "figure and ground" in a framework of stepping up the logical types as Bateson discussed. We did so through a workshop at an art museum held in a demonstrative manner. The details are described in the next section.

4   Workshop in an Art Museum

In this study, we would like to regard the arts as the imitation techniques used to identify the structure of the ambient space. Which is to say that it is not a problem of the technical transfer of skills acquired in one context to another. For example, studying painting will not lead to the making of a robot that can paint. However, creating a painting might utilize a modelling technique that has the same action level with producing robots or dramas, in terms of the modelling of the world.

This study accentuates the way of interdisciplinary research. Usually, as Bateson pointed out, interdisciplinary research occurs in a system in which, for example, an ecologist will need a geologist to tell him about the rocks and soils of the particular terrain which he is investigating (Bateson, 1972, p. 153). But in this study, we use interdisciplinary in another sense: for example, the man who studies the arrangement of leaves and branches in the growth of a plant finds the formal relations between stems, leaves, buds to be analogous to the formal relations of words in a sentence. He will discover a great academic value in such studies. Those who think first of the "things" which are related (the "relata") will dismiss any analogy between linguistic grammar and the anatomical structure of plants because they have no apparent similarity to each other. They will not accept any resemblance between a leaf and a noun. But, one way of interdisciplinary research is the very study of finding analogies between different things at a glance. In order to specify and step up the logical types in the invisible contexts, it is claimed that the forms of the relations between communication and the histogenesis process should be investigated. What should be imitated to reveal the formal relations is the form of the relationship structure. To conduct research against this background to investigate skill transference in art, a workshop was held in an art museum. Participants were directed to view some oil paintings and after that to copy them with pastel crayons, step by step. Visual artist Kenjiro Okazaki acted as facilitator. The details are as follows.

4.1   Time and Place

The workshop was held on December 24-25, 2011 in the exhibition hall and a large meeting room at the National Museum of Modern Art, Tokyo. The participants viewed the original paintings in the exhibition hall, and then moved to the large meeting room to reproduce the paintings which they viewed.

4.2   Participants

Fourteen adults participated (five men and nine women). The breakdown of their professions was as follows:

- one dancer
- one artist
- two researchers
- ten art students

4.3   Target Paintings for Reproduction

The target paintings were chosen with a focus on the historical development after the modern period, in which various stages of the creation process were reflected and observed in the style of painting; especially from the Impressionist period, through Constructivism, up to the abstract period. The selected paintings which the participants were required to reproduce in this workshop were the following five paintings:

- "Gold Necklace" by Ryuzaburo Umehara (1913)
- "Roses and a Girl" by Kaita Murayama (1917)
- "Renee, Green Harmony" by Henri Matisse (1923)
- "Water Mirror" by Yasuo Kazuki (1942)
- "Dog" by Kunitaro Suda (1950)

The selected paintings share a key feature in that the layout of the layers of colors perceived as figures and ground is crucial. For example, the paintings of

Matisse, Kazuki and Suda have the same typical layout, where the layer of the figure was first painted, and then marginal layers, perceived as background, were painted in the next step. Therefore, these paintings are comprised of a hind layer of the figure and a fore layer of the ground. For example, in the case of the small painting by Matisse (figure 1), the dark green portrait of a figure is surrounded by a light green background, which is perceived as showing a kind of depth such as falling into a dark hole.

Fig. 1. "Renee, Green Harmony" by Henri Matisse (1923)

4.4   Workshop Procedure

The facilitator, Okazaki, named each task from A to F (Table 1). In this study, we report on the results of these tasks (the reproduction of the paintings) except for tasks A-2, A-3, and B, which involved reproducing the artworks in words. We describe the directions that Okazaki gave the participants before they viewed the paintings in the exhibition hall as "memorizing." We categorized the directions given to the participants when they returned to the meeting room after viewing the paintings and before their reproducing task as "recall." The participants were seated in rows in the meeting room so as not to give them a chance to modify their works by referring to the reproduction processes of others. A video recording was made of all the participants' behaviors. Furthermore, in order to make the painting conditions the same for all participants they received the same type and number of drawing papers and the same set of pastel crayons (Sakura Craypas, 12 colors).

4.5   Criteria of Analysis

Six assessors were told to identify 14 types of behavior categories that we defined (Table 2), and they categorized the behaviors of the participants on the videos using the video analyzing software ELAN ver.4.1.2 at the maximum time resolution of the video rate. Each video scene was analyzed by two assessors.

4.6   Results and Discussion

Creating the Experience of Ambient Space by Using an Operating Surface Layout. The selected paintings in our workshop shared the common feature that the order of the layers between the figure and the ground is a significant key to understanding the paintings. Common sense dictates that usually the figure is regarded as being in front of the ground. But the selected paintings were created by using a reversal creation process, which was successful in inducing a strange perception. When we see the first painted surface layer in the center of the canvas, we perceive it as something like a depression, with depth. Conversely, in another painting we might perceive an ascending feeling. Using an operating surface layout creates the way the medium is perceived of the ambient space.

We are afraid to call the transition of perception induced by painting techniques the experience of beauty. To elaborate, we have doubts about whether the creation of beauty should be expected in the arts. To begin with, do artists pursue beauty? Akiba attempted to introduce Leibniz's idea of monadology to a new aesthetic apart from Kant's aesthetics (Akiba, 2011). Our question about art and beauty mentioned above corresponds with Akiba's thesis.

Kant attempted to give beauty the potential and the power to be constructed in order to overcome the problems (mutual constraints, that is, the antinomy) that are derived from each category of recognition (Kant, 1790). If this is so, however, we believe beauty gives us nothing but optimistic hopes that all will come out right, as if we are self-contained and without recognition of the truth of the world.

On the contrary, the monads are never optimistic. Following on from the philosophy of Aristotle, Leibniz postulated that sensing requires distance in time or space (that is, there is no direct perception of the object itself), and that we are surrounded by the medium (Leibniz, 1714). A simple explanation of monads is that the senses are confined by systems, but the outside is projected onto the inner perception. This means that the objects outside are understood only by a dissociation of the projected images between the inner perception and the objects outside. It was revealed in this workshop that our understanding of the perception of the object and depth, that is, the logical layers of space, was often reversed. We assume that we comprehend the very logical layers of the space (speaking in the Gibsonian viewpoint, the way of layout of surfaces); however, we can actually perceive only that one. Furthermore, what is perceived is the gap between steps of logical layers of the space which are indicated by color layers,

Table 1. Tasks and directions

Task A

Object	"Gold Necklace" by Ryuzaburo Umehara (1913) "Roses and a Girl" by Kaita Murayama (1917) "Renee, Green Harmony" by Henri Matisse (1923)
Memorization	"View the paintings for 15 minutes. Memorize the paintings as if you are witnesses."
Recall	"Reproduce the paintings in a limited time, six minutes for three paintings. Reduce the number of colors as much as possible. Draw with a simple combination of colors. You don't need to finalize it because these sketches are just memos."

Task C

Object	"Water Mirror" by Yasuo Kazuki (1942)
Memorization	"Take 40 seconds, just a short wink of time. Go and memorize the painting very quickly, and then draw the painting in full."
Recall	"Select one color, whichever you want. Reproduce the painting in full in your sketch. Take about 15 seconds. Draw your memorized impression quickly as if you are determining the layout of snapshots, and as if to fix the whole layout in one stroke and one touch. Say, 15 seconds is exaggerated, but as short a time as possible, in one minute is best." (During the reproducing process) "You can use the pastel crayons freely. Then you can paint the whole of all the pictures at the same time."

Task C2

Object	"Water Mirror" by Yasuo Kazuki (1942)
Memorization:	"One more time, you should go to see the paintings for 10 or 20 seconds. Reproduce the paintings in only two colors. Run to the hall to see the paintings and return here very quickly.In this case, three minutes for reproducing."
Recall	"You should not use two colors at the same time, nor use the two colors alternatively. If you begin to use the second color, you can't use the first color again. It's a two-color process like a color print. Never use the first color after you changed the color to the second."

Task D

Object	"Dog" by Kunitaro Suda (1950)
Memorization	"Same as the previous process. You should go to see the paintings for one minute and reproduce the paintings."
Recall	"Reproduce the paintings with only two colors. Just as in the previous process, if you begin to use the second color, you should not use the first color again. Yes, that way. Color over the drawing paper efficiently. You may guess the value of the two colors as if you are charged twice as much by a printer for using two colors."

Table 1. (*Continued.*)

Task E

Object	"Renee, Green Harmony" by Henri Matisse (1923)
Memorization	"Once more, go to see 'Renee' by Matisse. The time duration is much shorter than in the previous case, 30 or 50 seconds."
Recall	"Reproduce the paintings with only two colors. Follow the same method as the previous process in two minutes. The key point is which color to use first. You must meditate on which color should be first, considering that if you mistake the order of color use, you cannot repair it, resulting in crucial damage to the reproduction."

Task F

Object	"Gold Necklace" by Ryuzaburo Umehara (1913)
Memorization	"This painting's title, 'Gold Necklace', is significant. If you had to express this painting with one color, what is that color?" (In this case, the participants were required to reproduce the paintings without an opportunity to view the painting.)
Recall	"Use just two colors. Precise sketching ability is not required. Paint a simple figure of a human."

Table 2. Fourteen types of behavior categories

Behaviors without holding a crayon / Without contact with drawing paper

Pause: Without hand gesture	00
Movement: With hand gestures (painting gestures in air)	01
Looking for crayon	02

Behaviors with crayon

Drawing behaviors (Handling the crayon like a pencil / Contact with drawing paper)	
Tracing: Boundary (Tracing the outlines of the objects)	1
Reciprocating motion: Designated object (Drawing over the paper)	2
Reciprocating motion: Other objects (Drawing over the paper)	3
Painting behaviors (Drawing with the side of the crayon, like a wiper, or using small broken-off bits of crayon / Contact with drawing paper)	
Tracing: Boundary (Emphasizing the outline of the object such as brushing in white around the object)	4
Reciprocating motion: Designated object (Drawing over the paper)	5
Reciprocating motion: Other objects (Drawing over the paper)	6
Pause behaviors (Handling a crayon Without contact with the paper)	
Without hand gestures	P0
With hand gestures (painting gestures in air)	P1
Other (Rubbing the paper with fingers, etc.)	7

for example in the painting by Matisse, which demonstrate the breakdown of logic, or the gap as the immeasurable depth.

The object being perceived might be like a code that we provisionally presuppose in order to accept through making that breakdown of logic fit the framework. In this meaning, the object is regarded as an aim that perception processes cannot validate if it is not regarded as it is. Its existence is neither directly perceived nor proved definitely in advance but is regarded as the aim that makes the perception and action realized, and in the midst of the perception process this aim governs or regulates action, that is, the personal standard. For example, computing is only possible when the existence of a solution as well as the solvability of a problem is ensured. To the contrary, computing is meaningless when the solution is constantly valid. The thought and the perception are not established when the provisional object (referred to as "object X"), such as solutions in computing, may be derived but has not been directly comprehended. In other words, the object X is comprehended as what governs the medium, as the volume that parts recognition in the perception process; that is, the gap. Depth perception is the very possibility of action and does not enclose our perception in mere subjectivity. These are referred to as open access externalities, the conditions that enable the existence of others, for that reason. We suppose that the ambient space which Gibson defined can be interpreted in this way. The object X is the gap that enables our perception, the medium, and the opportunity for reflection. Additionally, that is, a personal standard, which is an opportunity for speculating about actions. The object X has not existed yet, but is definitely possible in the future. This temporal gap, the time delay, and the volume of the medium are the surrounding space of our perception, and this is also the condition of which time consists. This provisional object X can be something like a hope that is cast on the world, and is returned to us in the perception process.

We don't see the object in a painting, nor is the object painted in the painting. In discussing Matisse's painting, we first operate our perceptions, capture the gap that is the very depth (overlapping layers of paints), enter in it, and try to perceive the depth consciously; we perceive the image of the woman, resulting in finding out what the image is. A mystery of the existence of time and space that bundles and enables the relation between the object and the subject who sees, lies indeed in the time delay between the time we begin to glance at the paintings unmindfully and the time we find something there. Therefore, the paintings realize (more than express) the very time and space, that is, the medium. Otherwise, the paintings cannot express the space or the objects.

To reiterate, we substantively and concretely experience a void depth (a lack) that layers of colors reproduce. When our points of view (senses) enter that lack (the hole), devote themselves to it, and cannot escape from it, we can obtain acceptance in peace by naming it as "a woman" and recognize it again. We cannot deny this accepting process is not what is referred to as the perception of beauty. However, to name it "beauty" and make it completed in the way of pre-established harmony will prevent us from understanding the essential crisis of the process. The object is both assumed and broken down, therefore perception will become possible; that is, the production of admiration. Recognizing the clues of the depth or the gap of the internal perception to see the paintings without assurance of the object outside leads us to rediscover and recreate the object that is the image of the woman. It is indeed a real pleasure to admire these paintings. If we live in peace with the perception of beauty, comprehending real pleasure becomes difficult.

From Drawing to Painting. The understanding process of the workshop participants who were told to reproduce the paintings after viewing them has been estimated based on the video-recorded physical motions of the participants while reproducing the artworks. At the beginning of the workshop (task A), the participants received no constraints other than of time, and they reproduced the paintings with the free use of all twelve pastel crayon colors. In that period, they reproduced the paintings using many drawing techniques. To state it concretely, most of them tried to reproduce the object by demarcating outlines on the drawing paper. This means that most of the participants subjectively comprehended what the "object" is in the original paintings. But by the time for tasks C and D, Okazaki directed the participants to use only two colors, and furthermore he told participants to use the colors one after another in order, like a two-colored print. Around this time, most of the participants drastically changed their method of reproduction from a drawing-based method to a painting-based one, and the quality of their reproduced paintings obviously improved. One example is shown in figure 2.

Indirect Control as "Techne". In this workshop, the facilitator Okazaki did not give the participants direct-control directions: for example, "Do not use line drawing," "Do surface painting," and so on. It was found that the phase transition of behavior patterns occurred and produced the same structure of the original works when indirect-control directions were given. That is, the constraints on use of the painting tools rigidly restricted the temporal order of tool usage, and the directions functioned as boundary conditions for the participants' behaviors, leading to a desirable way of understanding the intentions of artists in order to capture the surrounding world. The distinction between direct and the indirect controls which is mentioned here comes from Deguchi's study which proposes the model of multi-subject complex systems (Deguchi, 2004). With direct control, the local constraints regulate the agent's behavior and the interactions between the agents for each. With indirect control, on the other hand, a global framework determines the global behavior of the whole system.

Giving explanations in advance about the creation process of the arts and direct specifications for the creation method seems as if it would be a more effective way to understand and create the arts. However, the results of this study suggested that indirect control is a more efficient way to provide constraints to people's unconsciousnesses than through other, direct ways. Rather than by transmitting directions directly to the consciousness, indirect control provides constraints to the unconscious self which are recognized and affirmed afterwards by the consciousness. In this way the body's behaviors can be manipulated.

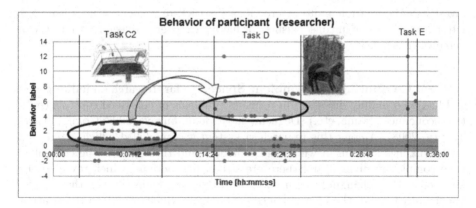

Fig. 2. Behaviors of the participants. Dark gray and light gray areas indicate drawing behaviors and painting behaviors, respectively.

Without an appeal to changing behavior at an unconscious level, directions encounter resistance from aspects such as the consciousness' self-awareness and self-respect, which does not lead to reflection on the behavior or a change in its direction.

Additionally, the reason that the people recognize indirect controls as "indirect" is that they are not aware of the body's consideration as "consideration", or regard it as involuntary. We can note that creators or those who can understand the creation process are aware that the most direct form of control in terms of skill transfer or understanding the works comes actually through an appeal to the body's consideration to indirect intelligence; in other words, direct perception. While this type of consciousness is peripheral and excluded from the point of view of the language center's functions, it is organized as a language in itself. This is in terms of "techne," the intelligence used to construct artifacts. The purpose of this workshop was to reveal empirically how "techne" is structured. If this is regarded as unconscious, it comes from suppressing consciousness that is governed by a special linguistic language.

In task B, early on between the tasks involving crayon reproductions of the paintings, the participants expressed their linguistic comprehension of the paintings (that is to say, to express the painting in words), which is not discussed here because the detailed analysis is not finished. Some participants who comprehended nothing but the outline of the object in the painting in the reproducing tasks nevertheless picked up the properties of the woman figure's depression as shown by using the word "pensive" for "Renee" by Matisse. That participant may have correctly perceived something at the unconscious level of sense, though they were not aware of the structure of paint layers on the canvas. An aim of this workshop was to specify the logical structure of ways to comprehend multilayered paintings, involving comprehension through the motions of reproduction and linguistic comprehension with free associations that seemed to have no relation to each other. In future, we have to develop more precise methods to investigate the invisible multi-layered logical structure without either flushing it into the black box of unconsciousness or having exaggerated feelings that we can perceive consciously.

Natural Computing and Ambient Space. So far, research on indirect control has been mainly discussed in terms of problems of contrived controls (politics, economics, drama, etc.) (Deguchi, 2004; Goan, Fukaya, 2007; Goan, Fukaya, Tsujita, 2008). Among this research, a system was modelled which acted as a dual dynamics system, in which the controller and the agents to be controlled were completely divided and there are strong interactions, which revealed the mechanism of exchanging information or of the state transitions in the system. On the other hand, the idea of contrived controls which adhere to the concept of "control," i.e. the concept that the controller should control other agents as it likes, is considerably weakened in the concept of natural computing proposed by Suzuki (Suzuki, 2012).

The following example explains this idea using the relationship between plants, harmful insects, and their natural enemies (Suzuki, Sakai, 2012). Plants (system) cannot and need not directly recognize both harmful insects and their natural enemies as objects. But the plant behaves (changes the conditions of the endocrine system) as if it can recognize insects and their natural enemies and control their behavior. What is the implementation mechanism for such plants' natural controlling behaviors? The purpose of their research is understood as clarifying the algorithm that induces state transitions in natural systems. Here we can note the possibility that the plant (system) can and may recognize the following two items, although they cannot directly recognize the existence or the behavior of harmful insects and natural enemies of the insects.

1. Recognition of the endocrine system's change: identify the trigger of the change and pay attention to the response relations.
2. Recognition of the fluctuations of the system's constancy (homeostasis): in other words, for example, perceive rhythmic change in physical conditions that get worse or become better.

A system involves these fluctuations and changes in a living process as part of its existence, even if it cannot directly recognize the objects outside the system; that is called "nature" in the concept of natural computing, and we call it "nature A." Therefore, when we assume a consciousness, the system becomes capable of being comprehended and measurable, even it is unconscious. That is like the weather condition.

Thus the ambient space can be understood as the space which enables the system to fluctuate as an adjusting process, or the field itself of which both measuring and adjusting simultaneously exist. Namely, that is the field or space itself in which a system, a species called plants in this example, can exist and be confined. In other words, the ambient space is the field that surrounds and confines the plants' intersubjective system, which is composed of surfaces and parameters that can be measured and extended. To make it possible for the plants to capture the critical discontinuities in the space – the change of physical conditions or the change of weather – is the very function of the ambient space. So the plants (system) become able to recognize the field on which they have dependence and in which they localize themselves, and in which they are able to exist. The field is comprised of the conditions that enable the plants' recognition, measurement, and computation of the field. Therefore the field enables the plants to measure and to compute the process of measurement or computing.

We call this field "nature X" to distinguish it from "nature A" previously mentioned. Nature X cannot be an "object" but be the "thing-in-itself." Nature X itself is the process of computing and recognition. It can be said that nature X is the recognition of the boundary between the inside (in which computing and the recursive repetition process are possible) and the outside (in which those are broken down). The preestablished harmony in recognition of the field induces no measurement or computing processes. The "repetition process" seems to repeat the same things, but actually it is not just repeating them. The significant part of the repetition process is to clarify the medium itself: the time flow whose homogeneity and continuity are never assured, in which it is possible for the computing process to induce the same solution every time. In other words, the importance of the computing process is not in the solution but in the recognition and the certainty that there are processes with no effect on the solution, and there are inevitable and irreversible distances. Not a computational solution, but the computing process itself is crucial to the medium property.

If we have an algebraic problem such as "what is one plus one?", the repetition process always induces the same solution, "two." But the short (or long) time lag for reaching the solution is the essential issue. The very time delay of this type produces the incident (or the art). This is not the issue with beauty. The concept of beauty is nothing but a convenient idea that finalizes the time flow, which is required to solve the puzzle. Due to indications that any difficult puzzle can be solved, it is easy to forget that there are valid distances, such as time flow and space.

Consider why the paintings do not consist of a unified algorithm. For example, the paintings differ from the input transfer, from the sensed information to the output, still picture. Why is this so? What does the painting of the space mean? What does the admiration of the paintings represent? What is the difference between the admiration of the paintings and the recognition of the paintings?

To address these questions, we would like to consider the following method, inspired by Aristotle's idea of entelekheia (Aristotle, 350BC): the significance of the computation (one's life) requests the computation (living) process. If we need only the results of cause and effect, the computing process may become redundant and could quickly be omitted. When the computation solutions are definite, the computation process is unnecessary. The process, the time delay, and the distance mentioned above are necessary for a slime mold, for example, but they have no meaning for observers who know the solution and regard knowing the solution as the computation. For those observers, the process of admiration becomes unnecessary. However, we emphasize that it is indispensable. This summarizes many of the significant issues explored in this workshop in the museum as well as Okazaki's goals (Okazaki, 2001). The concept of natural computing seems to involve the desire to avoid contrived control or the oppressive feeling from a regimented society. But even if this is so, the concept of computation tends to lead to a state that is called natural beauty, whether it is similar to that of Kantian philosophy or not. We fear that this tendency makes it difficult for us to recognize the medium property around the computation process; that is, the process itself, the distance, and the time flow. In order to avoid this tendency, we suggest that a future problem in the field of philosophy will be to bind an ethical stop standard to prevent it.

5   From the Integration of Multiple Viewpoints to the Dissection of Multi-logical Types and Asynchronous Cooperation

In this section, we would like to discuss the method of learning for those of us living in the ambient space. As mentioned at the beginning of this article, the incident that we call learning in this study is the step up of the logical type in the invisible meta-frame. We propose that the step up of the logical type happens when we accept that it is a dissection of the multi-logical types, deserting the idea that it is the integration of multiple viewpoints. The integration of multiple viewpoints is the idea that we cannot directly perceive the object itself, but we can reconstruct a unified image of the object using some or several numbers of partial perceptions. Namely, this means reconstruction of the unified object from multiple viewpoints. This also requires an assumption that there are some or many viewpoints to one object. As a one-to-many problem, this idea assumes the existence and the conditions of one (object X) from the beginning.

One difficulty with this idea is that it assumes the position (i.e., the unity) of the subject to the object. In this idea, however the viewpoints change, all the viewpoints are the viewpoints of the identified subject and it is tacitly assumed that their mutual continuity is assured. The position of the subject in the meta-level that integrates the multiple viewpoints is not considered to change. The viewpoints do not vary. There is no change in the relationship between the object and the subject that has an a priori assumption. Neither is there the definition nor the logical type of the space that surrounds them.

We suppose that this idea does not explain the learning mechanism. We insist that the logical type, including the relation between the subject and the object, must undergo changes or recomposition, not just a conversion of multiple viewpoints, or otherwise not only the relation between the subject and the object cannot change but also will not become stable. This is because a subject that integrates multiple viewpoints cannot avoid the infinite backward projection. The subject is always located outside the space. In other words, the subject of such a definition is not able to touch and to comprehend the ambient space that enables and ensures the relation between the subject and the object as a meta-context.

It is natural to consider that learning occurs when the transfer and a dissection of the multi-logical types happens, not the transfer and the integration of the multiple viewpoints. To elaborate, not only are the multi-logical types dissected but they also act in parallel and asynchronously, without being unified or integrated that is, they interact, relate to each other, shuffle the domain layers of each logic, and indirectly control each other. In other words, the step up of the logical type happens when lots of dissected individual logical domains (autonomous computing circuits, i.e., the algorithms) asynchronously cooperate. We have to consider that the subject exists only in ambient space. The production process and the meta-context, the meta-level and the ambient space are comprehensive only in the acting process.

Consider the admiration of the paintings. From the traditional view of "physical space" – that the object floats in an empty space– the paintings (and the space inside) and the subjects that admire them are distinguished from each other. Until now it has been thought that this space is what accepts "the subjects who admire the paintings (and the space inside)" and surrounds them. But adopting this idea makes it seem that the subject in such a space is fated to receive the inevitable infinite backward projection as mentioned above.

However, ambient space is quite a different definition of space from the traditional one. This associates with how the paintings represent and recognize the space surrounding the subjects who admire the paintings (and the space inside) where the figure-ground reversal perception cannot exist.

Before explaining the meaning of the previous sentence, we would now like to confirm the relation between the figure and the ground in the painting. Because the figure is the projection of the observer's differentiable codes, saying it is understood as "figure-ground reversal" is not correct, but "figure-figure switch" may rather correctly express the phenomenon. The strange phenomenon which occurs is that invisible things suddenly become recognized as figures. This is just a confusion of the observer's recognition circuitry. Similar to the territorial problem, it is nothing but the difference or fluctuation between the projection of the concept and the substance onto which the concept is projected. For other examples, it is similar to asking, "Is this a lizard or a gecko?" or "Is this a duck or a rabbit?" when we see an animal for the first time. As is well known, the duck-rabbit ambiguous image occurs in the natural world, therefore, these examples are more general phenomena than the problem of the paintings.

Learning in the ambient space is not a matter of "figure-ground reversal perception" but of "figure-figure switch." The parallelizing and recomposition of the basic logic is regarded as admirable in the paintings. For example, the ground emerges at the boundary between nature and artificiality, or between fortuity and inevitability. When this relation is reversed, this is the "ground-ground switch." This concept is associated with the realization of artificial intelligence or the artificial soul. The boundary between a human that has intelligence and a material body that has no intelligence becomes invalid, and their positions reverse, like a philosopher who, in contemplation of a desk, finds himself contemplated by the desk. The ground-ground switch also means that we may happen to detect the critical edge behind the ground.

Here, we should note that Bateson's learning theory, the step up of the logical types, and Gibson's learning theory of perception are coupled with each other. In our workshop at the art museum, we showed that there could exist a way of perceiving the ground-ground switch, the perception of surfaces' layouts, by stepping up from the logical type of figure-ground reversal perception – the figure-figure switch. It has been said that when we observe the figure, the ground (the background as the medium) is invisible. However, this is not true. On the contrary, we found that we can even perceive the depth at the interface of the ground-ground switch through the workshop experiences and the theory. This is an astonishing discovery that shows the possibility of finding out the logical crack for dissecting the world, leading to a radical change in our understanding of the relations of whole surfaces.

1. Breakthrough happens inside ourselves; breaking our eggshell, we will hatch in another world.
2. Visual sensation without vivid focusing is the very fundamental condition that the biological organ "retina" uses to capture the optical flow. This enables our body to comprehend the space that surrounds us and continues to move on.

Acknowledgement. This work was partially supported by a Grant-in-Aid for Scientific Research (B) No. 23320049 from the Japan Society for the Promotion of Science (JSPS). We are deeply grateful to Mr. Kenji Miwa, the curator of the National Museum of Modern Art, Tokyo for his great support in implementing the workshop.

References

Akiba, F.: Creating new aesthetics. Misuzu Shobo, Tokyo (2011)

Aristotle: De Anima (On the soul) (350 BC) (trans. by Nakahata, M.: Kyoto University Press, Kyoto (2001)

Bateson, G.: Steps to an ecology of mind. The University of Chicago Press, Chicago (1972)

Deguchi, H.: Economics as an agent-based complex system. Springer-Verlag Tokyo, Tokyo (2004)

Gibson, J.J.: The perception of the visual world. Houghton Mifflin, Boston (1950)

Gibson, J.J.: The ecological approach to visual perception. Houghton Mifflin, Boston (1979) (republished in 1986 from Lawrence Erlbaum Associates, Hillsdale)

Goan, M., Fukaya, T.: Timing makes meaning: Protocol analysis on directions during creation in a drama-making process. In: Proc. of International Symposium on Skill Science 2007 (ISSS 2007), pp. 68–75 (2007)

Goan, M., Fukaya, T., Tsujita, K.: Two Succeeding Stages in Acquisition Process of a Rehearsed Drama: Applying System Dynamics to Human Collaborative Behavior. In: The 2nd International Conference on Knowledge Generation, Communication and Management (KGCM 2008), Proc. of The 12th World Multi-Conference on Systemics, Cybernetics and Informatics (WMSCI 2008), vol. VII, pp. 126–131 (2008)

Kant, I.: Kritik der Urteilskraft (1790) (trans. by Sakata, T.: Kawade Shobo Shinsha, Tokyo (1973))

Leibniz, G.W.: La Monadologie (1714) (trans. by Nishitani, Y., Yoneyama, M., Sasaki, Y.: Kousakusha, Tokyo (1989))

Lombardo, T.J.: The reciprocity of perceiver and environment. Lawrence Erlbaum Associates, NJ (1987)

Okazaki, K.: Inhabitants of the moon. 10+1, vol. 23. INAX Press, Tokyo (2001) (in Japanese), http://www.eris.ais.ne.jp/~fralippo/module/Study/OKK030811_probability1/index.html

Rubin, E.: Synsoplevede Figurer. Gyldenhal, Copenhagen (1915)

Suzuki, Y.: Introduction. Abstracts of the 6th International Workshop on Natural Computing (IWNC6) (2012)

Suzuki, Y., Sakai, M.: Co-Evolution of info-chemical signal in Chemical Ecology. Abstracts of the 6th International Workshop on Natural Computing (IWNC6) (2012)

Taylor, R.P., Micolich, A.P., Jonas, D.: Using science to investigate Jackson Pollock's drip paintings. Journal of Consciousness Studies 7, 137–150 (2000)

Aesthetic Aspects of Technology-Mediated Self-awareness Experiences

Junji Watanabe

NTT Communication Science Laboratories,
Nippon Telegraph and Telephone Corporation, Atsugi, Japan
watanabe.junji@lab.ntt.co.jp

Abstract. Today's media technologies can provide experiences where people become aware of their fundamental attributes and unconscious behaviors. In this paper, I introduce three interface technologies designed to mediate such self-awareness experiences, and interpret them from an aesthetical viewpoint.

Keywords: Self-awareness experiences, Interface technologies, Aesthetics, Saccade-based display, Galvanic vestibular stimulation, Heartbeat picnic.

1 Introduction

Technologies have been tremendously important for expanding our sphere of existence and solving problems. In the years ahead, I envision that technologies will play an important role in probing the depths and opening up new vistas of the human mind. Interface technologies available in today's modern society can provide experiences where people become aware of fundamental attributes and unconscious behaviors to which they normally pay little attention as they lead their lives. My colleagues and I have developed unique interface technologies based on the perceptual characteristics of humans and have had opportunities to introduce them to a wide range of people in the technology demos (e.g., [1-4]), presentations at science museums (e.g., [5-6]), and art festivals (e.g., [7-9]). Through these exhibitions, I realized that experiences mediated by the interface technologies can promote people recognize self and aware of new rules as to how they relate to their environments.

The self-awareness experience and the aesthetic judgment defined by Immanuel Kant are analogous in that both are based on subjective appreciation of a phenomenon and implication of its universality. This is because the aesthetic judgment is related to the senses of being "purposive without a specific purpose" (a foundation of accepting existence of a specific phenomenon) and "lawful without a specific law" (a foundation of supporting its universality) [10]. Here I conjecture that the self-awareness experiences by the technologies can be a way of understanding ourselves by extending subjective experience to others (aesthetic approach), without assuming objective observation (scientific approach). In this paper, I introduce three specific themes of the self-awareness experiences motivated by the aesthetic approach: themes about constraints of perception (section 2), behavior (section 3), and body (section 4).

Y. Suzuki and T. Nakagaki (Eds.): WSH 2011 and IWNC 2012, PICT 6, pp. 148–153, 2013.

2   See What You Throw off from Visual Perception

It is well known that humans actively collect information from the outside world, but this inevitably means that they also discard information. Consider vision, for example. It is known that when one looks in the direction one wants to see, the eye is moved very rapidly, but one will not perceive any movement of visual images. The visual information during rapid eye movements is discarded in order to build up a fixed stable visual scene. This capability of the brain to construct and maintain "a stable world image" is critical for our everyday lives.

Using an interface technology, we can get a real sense of how we construct stable world images. In the interface technology, I and my colleague (H. Ando) and supervisors (T. Maeda and S. Tachi) focused on the information that is discarded and leveraged this information to develop a novel visual information presentation method [2, 11, 12]. To describe the principle of the technology, let us first consider the scheme illustrated in Fig. 1(a) consisting of a one-dimensional array of lights that moves rapidly back and forth to produce a two-dimensional (2D) visual image. This type of visual display is already available on the market (e.g., [13]). An alternative scheme is illustrated in Fig. 1(b). In this scheme, which we call "Saccade-based Displays," the array of lights does not move but an observer's eye movement called a saccade is made in front of rapidly flashing lights. The one-dimensional flashing pattern during the eye movement is projected onto different locations of the retina, and these projections create a coherent 2D image momentarily. Visual displays based on this perceptual phenomenon can present full life-size color visual images without a projection screen (even in midair) using only a single column of light (Figs. 2), and have been used in the field of arts and entertainments [6, 14, 15].

Generally, the selective culling process of visual images does not raise to the level of consciousness, but in our information presentation scheme, the act of seeing itself is experienced as one's own personal visual image. This experience makes people aware of their own eye movements, which they had never been conscious of before. This calls attention to how we perceive the world and how much information we discard.

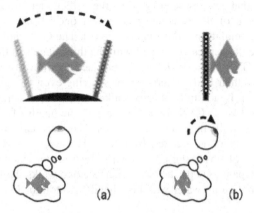

Fig. 1. Schematic drawing of the information presentation principle using a moving light column (a) and that using the observer's eye movements (b)

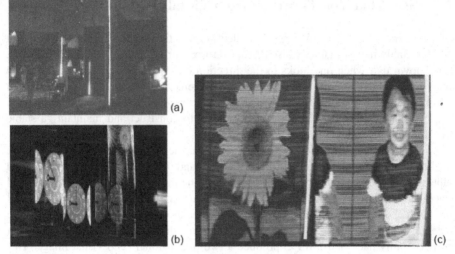

Fig. 2. Perceived image with observer's eyes static (a). Two LED columns 1.8 m in height and one LED column 4 m in height are shown in this photo. An example of perceived images displayed with Saccade-based Display (b). Life-size photographic images can be presented (c). The images are 128 pixels in vertical and horizontal resolutions with 4-bit color depth. Images (b) and (c) were taken with a slow-shutter rotating camera.

3   Save Yourself from Constraints of Free Will

With recent advances in personal electronic devices, ubiquitous sensors, and high-speed networks, technologies have become transparent, and they affect, and occasionally control, us without our noticing. Under these circumstances, both attributing everything to one's free will and to being forced by the outside world are problematic. From this perspective, I and my colleagues (H. Ando and T. Yoshida) have devised a scheme, called "Save YourSelf !!!" (Fig. 3), which can create a physical experience that enables one to sense "free will" under unseen control [7].

The basic principle of the scheme is to distort one's sense of balance using a galvanic vestibular stimulation (GVS) interface [16]. The GVS interface (a headset in Fig. 3) can stimulate the nerves associated with the three canals in the inner ear with a weak current. The GVS causes a lateral virtual acceleration toward the anode and makes users feel like their body is lurching or swaying even though they may not be actually moving at all. In addition, as they walk it can induce lateral divergence from the intended straight line. If GVS control is put into the hands of another person, the user can be made to walk in a certain direction. This scenario reflects a situation where a person's behavior is dictated by others. However, "Save YourSelf !!!" is based on the scheme of observing one's own state from an external perspective. In this scheme, a participant puts on the GVS interface and walks freely while holding a clear plastic bowl containing water with both hands. A monitor-like figurine floats on

Fig. 3. Photograph of "Save YourSelf !!!" (© Tomofumi Yoshida)

the water in the bowl. The movement of the figurine is linked to the participant's sense of balance since the data obtained by a direction sensor embedded in the figurine is sent to the GVS interface. The participant walks while keeping the balance of the figurine floating on the water. In this experience the relation between the self and the outside world is externalized, and observing the figurine floating on the water (a metaphor of the externalized self) has participants realize the nature of self-identity in today's life.

4   Real Sense of Life Generated by Touching Heartbeat

Although we do become conscious of the body under certain circumstances, (e.g., when we catch a cold), we are oblivious to the body of self. For example, although we know that the heart is indispensable for every human being, we lead our lives without paying much attention to it. Here, for a workshop called "Heartbeat Picnic," I and my colleagues (Y. Kawaguchi, K. Sakakura and H. Ando) developed a simple device as shown in Fig. 4(a) for reaffirming people's lives [8]. The participants were asked to hold a stethoscope in one hand and a vibration speaker (referred to as the heart box) in the other as shown in Fig. 4(b). When the stethoscope was placed on their chest, their own heartbeat was output as both sounds and vibrations from the heart box, which enabled them to not only hear their own heartbeats but also feel them as vibrations. Since the workshop was held outdoors, they were free to move about as if they were having a picnic, and they were able to feel the changes in their heartbeat with their own hands. Moreover, as shown in Fig. 4(c), by exchanging their heart boxes with those of other participants, they were able to feel the differences between their own heartbeats and those of others. The experience of touching heartbeats, which is usually impossible, provided the participants with the opportunity to appreciate the importance of his/her own life as well as the lives of others.

Life appreciation workshops are usually held in a natural environment where participants can directly experience the abundant life. In every day life, however, there are almost no opportunities for us to appreciate our own lives. This workshop is aimed at enabling participants to experience the reality of life by evoking the imagination through the sense of touch rather than having to escape from their modern everyday environment. I believe that the sense of touch enables us to have a physical and realistic connection with the wealth of our lives.

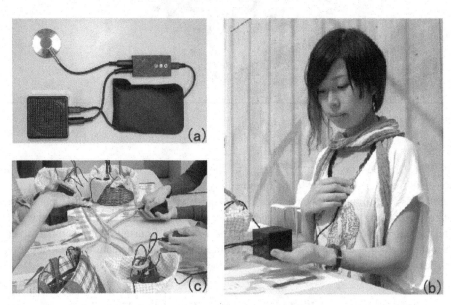

Fig. 4. Photographs of "Heartbeat Picnic." Equipments used in the workshop are stethoscope, vibration speaker (heart box), signal processing circuit, and batteries (a). Equipment in use (b). Exchange of heart boxes among participants (c)

5   Conclusion

One doesn't know the "self" initially, but only gradually becomes acquainted with it as it is manifested though interaction with the environment and others. The interface technologies described above can provide unique experiences of self with implications of the common features of human beings. I hope that feeling unconscious characteristics of ourselves broadens perspectives of ourselves and frees ourselves from the constraints. Although the notion of applying deep insight into the aspects of "the self" may strike one as the realm of philosophical or art practices, it has important implications for everyday living and an affinity for interpretation from the viewpoint of aesthetics.

References

1. Ando, H., Amemiya, T., Maeda, T., Nakatani, M., Watanabe, J.: Embossed Touch Display: Illusory Elongation and Shrinking of Tactile Objects. In: Emerging Technologies, SIGGRAPH 2006, Boston, U.S.A. (2006)
2. Ando, H., Watanabe, J., Amemiya, T., Maeda, T.: Full-scale Saccade-based Display: Public/Private Image Presentation based on Gaze-Contingent Visual Illusion. In: Emerging Technologies, SIGGRAPH 2007, San Diego, U.S.A. (2007)
3. Ooshima, S., Fukuzawa, Y., Hashimoto, Y., Ando, H., Watanabe, J., Kajimoto, H.: /ed - Gut Feelings when Being Cut and Pierced -. In: New Tech Demo, SIGGRAPH 2008, Los Angels, U.S.A. (2008)
4. Watanabe, J., Kusachi, E., NOSIGNER, Ando, H.: Touch the Invisibles. In: Information Aesthetics Showcase, SIGGRAPH 2009, New Orleans, U.S.A. (2009)
5. Ando, H., Watanabe, J.: Sensory Circuit Collection. In: Device Art Collection. National Museum of Emerging Science and Innovation (MIRAIKAN) (2009)
6. Ando, H., Watanabe, J., Tabata, T., Verdaasdonk, M.A.: Saccade-based Display –blink to see __-. In: Device Art Collection. National Museum of Emerging Science and Innovation (MIRAIKAN) (2011), http://www.junji.org/saccade/
7. Ando, H., Yoshida, T., Watanabe J.: Save YourSelf !!! Ars Electonica Center, Linz Austria (2007), http://www.junji.org/saveyourself/
8. Watanabe, J., Kawaguchi, Y., Sakakura, K., Ando H.: Heartbeat Picnic. Ars Electronica Exhibition, OKcenter, Linz, Austria (2011), http://www.junji.org/heartbeatpicnic/
9. Watanabe, J., Kusachi, E., Ando, H.: Touch the Invisibles. In: 12th Japan Media Arts Festival, Roppongi, Tokyo, Japan (2009)
10. Kant, I.: The Critique of the Power of Judgment (1790)
11. Watanabe, J., Ando, H., Maeda, T., Tachi, S.: Gaze-contingent Visual Presentation based on Remote Saccade Detection. Presence: Teleoperators and Virtual Environments 16(2), 224–234 (2007)
12. Watanabe, J., Maeda, T., Ando, H.: Gaze-contingent Information Display with Electro-oculo-graph-based Saccade Detection. ACM Trans. Applied Percept 9(2), 1–12 (2012)
13. Dynascan Techinology Inc. (2006), http://www.dynascanusa.com/
14. Bell, B., Parks, T.E., Post, R.B.: Elusive Imagery of the LightStick. Leonardo 19(1), 3–10 (1986), See also the artwork by Bill Bell at his website "Subliminary Artworks" http://www.subliminaryartworks.com
15. Watanabe, J., Tavata, T., Verdaasdonk, M.A., Ando, H., Maeda, T., Tachi, S.: Illusory Interactive Performance by Self Eye Movement. Sketches. In: SIGGRAPH 2004, Los Angeles, U.S.A. (2004)
16. Maeda, T., Ando, H., Amemiya, T., Nagaya, N., Sugimoto, M., Inami, M.: Shaking the World: Galvanic Vestibular Stimulation as a Novel Sensation Interface. Emerging Technologies. In: SIGGRAPH 2005, Los Angels, U.S.A. (2005)

Author Index

Akiba, Fuminori   71, 117, 119

Fricker, Mark D.   14
Fujiwara, Yoshi   30

Goan, Miki   130

Hagiya, Masami   39

Iba, Hitoshi   82, 93
Ishikawa, Takuma   130
Ito, Kentaro   14

Kawamata, Ibuki   39
Kihara, Susumu   130
Kunita, Itsuki   3, 14

Lee, Chiu Fan   14

Nakagaki, Toshiyuki   3, 14
Noman, Nasimul   82, 93

Okazaki, Kenjiro   130

Palafox, Leon   82, 93

Saigusa, Tetsu   3
Sato, Sho   3
Sekine, Ryoji   104
Suzuki, Yasuhiro   49

Takashima, Shinichi   130
Tero, Atsushi   14
Tsujita, Katsuyoshi   130

Watanabe, Junji   148

Yamamura, Masayuki   104
Yoshihara, Kazunori   14